Artificial Intelligence and Human Performance in Transportation

Artificial intelligence (AI) is a major technological advance in the 21st century. With its influence spreading to all aspects of our lives and the engineering sector, establishing well-defined objectives is crucial for successfully integrating AI in the field of transportation. This book presents different ways of adopting emerging technologies in transportation operations, including security, safety, online training, and autonomous vehicle operations on land, sea, and air.

This guide is a dynamic resource for senior management and decision-makers, with essential practical advice distilled from the expertise of specialists in the field. It addresses the most critical issues facing transportation service providers in adopting AI and investigates the relationship between the human operator and the technology to navigate what is and is not feasible or impossible. Case studies of actual implementation provide context for common scenarios in the transportation sector.

This book will serve the reader as the starting point for practical questions regarding the deployment and safety assurance of new and emergent technologies in the transportation domain. **Artificial Intelligence and Human Performance in Transportation** is a thoughtful and enlightening read for professionals in the fields of Human Factors, Engineering (Aviation, Maritime, and Land), Logistics, Manufacturing, Accident Investigation and Safety, Cybersecurity, and Human Resources.

Artificial Intelligence and Human Performance in Transportation
Applications, Challenges, and Future Directions

Edited by
Dimitrios Ziakkas and Anastasios Plioutsias

CRC Press is an imprint of the
Taylor & Francis Group, an **informa** business

Front cover image: Irina Shi/Shutterstock

First edition published 2025
by CRC Press
2385 NW Executive Center Drive, Suite 320, Boca Raton FL 33431

and by CRC Press
4 Park Square, Milton Park, Abingdon, Oxon, OX14 4RN

CRC Press is an imprint of Taylor & Francis Group, LLC

© 2025 Dimitrios Ziakkas and Anastasios Plioutsias

Reasonable efforts have been made to publish reliable data and information, but the author and publisher cannot assume responsibility for the validity of all materials or the consequences of their use. The authors and publishers have attempted to trace the copyright holders of all material reproduced in this publication and apologize to copyright holders if permission to publish in this form has not been obtained. If any copyright material has not been acknowledged please write and let us know so we may rectify in any future reprint.

Except as permitted under U.S. Copyright Law, no part of this book may be reprinted, reproduced, transmitted, or utilized in any form by any electronic, mechanical, or other means, now known or hereafter invented, including photocopying, microfilming, and recording, or in any information storage or retrieval system, without written permission from the publishers.

For permission to photocopy or use material electronically from this work, access www.copyright.com or contact the Copyright Clearance Center, Inc. (CCC), 222 Rosewood Drive, Danvers, MA 01923, 978-750-8400. For works that are not available on CCC please contact mpkbookspermissions@tandf.co.uk

Trademark notice: Product or corporate names may be trademarks or registered trademarks and are used only for identification and explanation without intent to infringe.

Library of Congress Cataloging-in-Publication Data
Names: Ziakkas, Dimitrios, editor. | Plioutsias, Anastasios, editor.
Title: Artificial intelligence and human performance in transportation : applications, challenges, and future directions / edited by Dimitrios Ziakkas and Anastasios Plioutsias.
Description: First edition. | Boca Raton, FL : CRC Press, 2025. | Includes bibliographical references and index. |
Identifiers: LCCN 2024015831 (print) | LCCN 2024015832 (ebook) |
ISBN 9781032751986 (hbk) | ISBN 9781032770222 (pbk) | ISBN 9781003480891

ISBN: 978-1-032-75198-6 (hbk)
ISBN: 978-1-032-77022-2 (pbk)
ISBN: 978-1-003-48089-1 (ebk)

DOI: 10.1201/9781003480891

Typeset in Times
by codeMantra

To our families, for their words of encouragement and understanding despite the many hours we devoted to this work. Thank you for your dedication, invaluable contributions, and the unwavering support provided while bringing this project to life.

Contents

Preface xviii
Foreword xix
Acknowledgments xxiii
About the Editors xxiv
Contributors xxvi

Introduction 1

Chapter 1 Artificial Intelligence in Aviation 3

    1.1   Conceptual Framework for Artificial Intelligence in
         Transportation 3
         Konstantinos Sakalis

    References 5

    1.2   Evolution of AI and Its Potential Applications 5
         Konstantinos Sakalis

    References 7

    1.3   Benefits of AI Implementation in Transportation 7
         Konstantinos Sakalis

         1.3.1   AI on Transforming Transportation 7
    References 8

    1.4   Control Ergonomics in the Era of AI 9
         Anastasios Plioutsias and Dimitrios Ziakkas

    References 10

    1.5   Learning to Trust AI and ML – Key Steps for
         Implementing Changes 11
         Dimitrios Ziakkas and Michail Diakomihalis

    References 14

    1.6   AI Certification & Challenges 15
         Georges Rebender

         1.6.1   Why Should We Regulate? 15
         1.6.2   AI Regulatory Approaches 16
         1.6.3   EU Aviation AI Blended Approach 19
         1.6.4   Conclusions 23
    References 23

1.7 Simulated ATC Environment (SATCE) and ASTi SERA's Role in CBTA: The Purdue University Case Study 24
Dimitrios Ziakkas

Reference 25

Chapter 2 Artificial Intelligence Training and Operations 26

2.1 Incorporating AI into Transportation Higher Education: Curriculum Considerations for an Evolving Industry 26
Debra Henneberry and Julius Keller

    2.1.1 Contemporary Examples of Artificial Intelligence Being Integrated into Higher Education 29
    2.1.2 Summary 30

References 30

2.2 The Role of AI in AI Training/Operations 31
Dimitrios Ziakkas, Konstantinos Pechlivanis, and Anastasia Kosmidou

    2.2.1 Mental Rehearsal in Aviation and Emerging Technologies: The Step Forward in Training 31
    2.2.2 Benefits of Extended Reality (XR) in Pilots' Mental Rehearsal 32
    2.2.3 Artificial Intelligence in Shipping Operations 33
Michail Diakomihalis

References 34

2.3 The Role of AI in Weather Prediction, Planning, Route Optimization, and Scheduling 35
Anastasios Plioutsias and Dimitrios Ziakkas

    2.3.1 AI in Weather Prediction 35
    2.3.2 AI in Planning and Decision Making 35
    2.3.3 Route Optimization 36
    2.3.4 Scheduling Adjustments 37
    2.3.5 AI in Weather Prediction, Planning, Route Optimization, and Scheduling: The Shipping Case Study 37

References 38

2.4 AI-Powered Maintenance and Predictive Analysis for Vehicle Health Monitoring and Maintenance Scheduling 38
Ioannis Katsidimas and Athanasios Kotzakolios

    2.4.1 Big Data Analytics in Aviation Maintenance 39
    2.4.2 Machine Learning for Failure Prognostics 40

		2.4.3	Integrated Frameworks and Data-Driven Techniques for Predictive Maintenance 41
		2.4.4	Digital Twins 43
	References 44
	2.5	Artificial Intelligence Applications as Cognitive Artifacts in Aviation Distributed Cognition 45
		Konstantinos Pechlivanis
	References 46
	2.6	The VTR Case Study...46
		Evey Cormican
		2.6.1	The Genesis of Visionary Training Resources (VTR)...46
		2.6.2	VTR Products Show Positive Results in Proof-of-Concept (POC) 47
		2.6.3	Incorporating VR into Pilot Training 47
		2.6.4	Generating and Using VR Data................................47
		2.6.5	The Future of Virtual Reality Airline Training 48

Chapter 3 Artificial Intelligence in Traffic Management 49

	3.1	Applications of AI in Traffic Management and its Challenges..49
		Lea-Sophie Vink
		3.1.1	Individual Vehicles – Challenges and Benefits 49
		3.1.2	Network and Traffic Management.............................51
		3.1.3	Merging of Two Directions 52
	References 52
	3.2	Traffic Prediction and Optimization 53
		Anastasios Plioutsias and Dimitrios Ziakkas
		3.2.1	Traffic Optimization...53
		3.2.2	Future Directions in AI for Aviation Traffic Management 53
		3.2.3	Safety and Security 54
		3.2.4	Technological Challenges in AI for Aviation Traffic Management 54
	References 55
	3.3	Role of AI in Unmanned Aircraft Systems and the Drone Traffic Management System..55
		Andrew Black
		3.3.1	Augmenting the Envelope: Machine Assistance to Remote Aviation (MARA)..56

	3.3.2	What Is Machine Assistance? 56
	3.3.3	MARA and EASA 57
	3.3.4	Future Research..57
	3.3.5	Conclusion 57

References 58

3.4 Role of AI in Unmanned Aircraft Systems and the Drone
 Traffic Management System 58
 Andrew Black

	3.4.1	The World below 400 Feet Above Ground Level..58
	3.4.2	Towards Control in the Uncontrolled 59
	3.4.3	MARA Monitoring 59
	3.4.4	Conclusions 59

References 60

3.5 The Role of AI in Training of Traffic
 Controllers – Managers 60
 Lea-Sophie Vink

	3.5.1	Overview of the Typical ATCO Training Cycle 60
	3.5.2	Where AI Can Be Used to Improve and Speed Up Training 61
	3.5.3	Conclusions 61

References 62

3.6 Monitoring Human Performance, Safety Intelligence, and
 Fatigue Risk Management in Traffic Ecosystems 62
 Spyridon Chazapis

	3.6.1	Fatigue Risk and Workload Management Applications of Artificial Intelligence.........................62

References 63

3.7 Safety Assurance, Safety Management, and the Design of
 Airspace/Traffic Management... 64
 Lea-Sophie Vink

	3.7.1	Human Performance and Fatigue as Examples of Weak Signal Analysis... 64
	3.7.2	Why AI Will Take This and Make It Better, but with Challenges 64
	3.7.3	How Does Safety Intelligence Feed Safety Assurance?... 64
	3.7.4	Next Steps Towards Parameterization......................... 65

References 65

Contents

3.8 The Austro Control Case Study 65
Lea-Sophie Vink

- 3.8.1 The Components and Master Supervisory Decision-Support System 66
- 3.8.2 Evolving to an AI Driven System with Humans in the Loop 67
- 3.8.3 Conclusions and Building Trust 68

Chapter 4 Artificial Intelligence in Airport – Ports – Train Stations Operations 69

4.1 Applications of AI in Airports – Ports – Train Stations Management, Security, and Surveillance Systems 69
Dimitrios Ziakkas

- 4.1.1 Introduction 69
- 4.1.2 Airports 69
- 4.1.3 Ports 69
- 4.1.4 Train Stations 70
- 4.1.5 The Shipping Companies Case Study 70

Michail Diakomihalis

References 71

4.2 The Use of AI in Baggage Handling, Passenger Screening, and Facial Recognition within Airport, Port, and Train Station Operations 71
Anastasios Plioutsias

- 4.2.1 AI in Passenger Screening 72
- 4.2.2 AI in Facial Recognition 72
- 4.2.3 Integration Challenges and Solutions 72

References 74

4.3 The Role of AI in Multiple Remote Towers and Ground Operations 74
Stefano Conte

- 4.3.1 AI Solutions for Multiple Remote Tower Operations 74
- 4.3.2 AI Solutions for MRTO Challenges 75
- 4.3.3 Conclusion 77

References 77

4.4 Artificial Intelligence in Shipping Operations 77
Michail Diakomihalis

- 4.4.1 Future Options 79

References 80

4.5 Train Operations 80
Beatrix Walzl

    4.5.1 Applications of AI in Train Operations 80
    4.5.2 Challenges in Implementing AI in
           Train Operations 81
    4.5.3 Outlook and Conclusions 81
References 82

Chapter 5 Artificial Intelligence in Customers' Experience 83

5.1 AI-Based Chatbots and Virtual Assistants for
Personalized Customer Interactions 83
Anastasios Plioutsias and Dimitrios Ziakkas

    5.1.1 Evolution of Chatbots and Virtual Assistants 83
    5.1.2 Technologies Behind AI Chatbots and Virtual
           Assistants 83
    5.1.3 Applications in Customer Service 84
    5.1.4 Benefits for Businesses and the
           Aviation Industry 84
    5.1.5 Challenges and Considerations 84
    5.1.6 Future Trends 84
References 84

5.2 Use of AI Recommendation Systems, Pricing
Optimization, and Revenue Management 85
Volodymyr Bilotkach

    5.2.1 Introduction 85
    5.2.2 Opportunities for Traditional RM 86
    5.2.3 Optimizing Add-On Pricing and Sales 86
    5.2.4 Using AI for Sales of Commission-Based
           Products and Advertising 87
    5.2.5 Getting more Out of FFPs 87
Reference 88

5.3 The Role of AI in Customer Feedback Analysis,
Sentiment Analysis, and Social Media Monitoring 88
Caterina Ciacatani

    5.3.1 Customer Feedback Analysis 88
    5.3.2 Sentiment Analysis 89
    5.3.3 Social Media Monitoring 90
    5.3.4 Conclusion 90
References 90

5.4 The Airliners' Case Study 91
Anastasios Plioutsias and Dimitrios Ziakkas

Contents   xiii

	5.4.1	Background 91
	5.4.2	Technical Integration with Legacy Systems 91
	5.4.3	Outcomes... 91
Reference 92

Chapter 6 Artificial Intelligence in Maintenance, Technical Support and Repair Operations 93

  6.1 Role of AI in Predictive Maintenance and Condition-Based (CB) Monitoring for Vehicle Systems 93
       Ioannis Bakopoulos

       6.1.1 Incorporating AI into Aircraft and Maintenance – Documentation and Sensor Data 93
       6.1.2 Supply Chain Management 95
       6.1.3 Conclusion 96
       References 96

  6.2 Use of AI in Fault Detection, Diagnosis, and Prognosis for Improved Maintenance Efficiency 97
       Timothy D. Ropp

       6.2.1 Fault Detection 97
       6.2.2 Fault Diagnosis 99
       6.2.3 Fault Prognosis – Predicting Failure and Prescribing Fixes 100
       References 100

  6.3 AI Applications in Repair Operations and Spare Parts Management 101
       Konstantina Ioannidi

       6.3.1 AI in the Aviation Industry (Case Study) 102
       6.3.2 AI in Rail Transportation (Case Study) 103
       6.3.3 AI in the Shipping Industry (Case Study) 104
       References 104

  6.4 AI in Maintenance and Supply Chain 105
       Konstantina Ioannidi

       References 106

  6.5 AI Uses for Technical Support in Transportation Management - Monitoring of Systems: Automation of the Radio PIREP Submission Process in General Aviation (GA) 107
       Shantanu Gupta

       References 107

6.6	Data Governance: The Basis for AI in Maintenance and Repair Operations .... 108 *Timothy D. Ropp*	
	6.6.1 AI and MRO Business Metrics Considerations 110	
	6.6.2 Predictive Maintenance and Its Tools in MRO 110	
References 111		
6.7	The Purdue University School of Aviation and Transportation Technology (SATT) Case Study .... 112 *Timothy D. Ropp*	
	6.7.1 Digital Twin Tools in MRO 113	
Reference 114		

Chapter 7 Human-Machine Interaction in Artificial Intelligence in Aviation 115

7.1 Human Factor Considerations in the Implementation of AI in Transportation 115
Anastasios Plioutsias and Dimitrios Ziakkas

    7.1.1 Key Human Factor Considerations 115
    7.1.2 Challenges and Opportunities 116
    7.1.3 Case Study 116

References 116

7.2 Training and Skill Development for Human Operators Working with AI Systems 117
Anastasios Plioutsias and Dimitrios Ziakkas

    7.2.1 Understanding AI and Its Role in Transportation 117
    7.2.2 Core Skills for Human Operators 117

References 118

    7.2.3 Prompt Engineering in Generative AI Applications for Transportation: An Emerging Competency 118
Konstantinos Pechlivanis

References 119

7.3 The Future of Simulation in HMI in AI in Transportation 119
Anastasios Plioutsias and Dimitrios Ziakkas

    7.3.1 Implications for HMI 120
    7.3.2 Future Trends 120

References 120

7.4 Automation Framework for Human Decision-Making and Key Uses of AI for Cognitive Information Processing 121
Shantanu Gupta

    7.4.1 Automation in Decision Making 121
    7.4.2 Types of Decision Automation 121

References 122

7.5 Adaptive Automation and AI in Predictive Human Performance 122
Shantanu Gupta

    7.5.1 Adaptive Automation Framework 123

References 124

7.6 Human-AI Teaming by Implementing AI-Based Applications on Connected Electronic Transportation Devices 124
Shuo Liu

    7.6.1 Trajectory-Based Operations 124
    7.6.2 Air Traffic Control Loop 125
    7.6.3 Operation Considerations 125

References 127

7.7 The Electronic Flight Bag Case Study: Exploring the Potential of Machine Learning in 4D Operations 127
Shuo Liu

Chapter 8 Ethical and Legal Challenges in the Implementation of Artificial Intelligence 128

8.1 Ethical Considerations in the Responsible Implementation of AI in Transportation 128
    8.1.1 AI Ethics in Aviation 128
Konstantinos Pechlivanis

References 129

    8.1.2 Ethical and Legal Considerations in Responsible AI Implementation in Shipping 129
Michail Diakomihalis

References 130

8.2 Legal Frameworks and Considerations in the Responsible Implementation of AI in Transportation 131
Thanasis Binis, Anastasios Plioutsias, and Dimitrios Ziakkas

		8.2.1	Autonomous Vehicles (AVS) 131
		8.2.2	Data Protection and Privacy 131
		8.2.3	Regulatory Compliance and Standardization 131
		8.2.4	Conclusion 131
	References 132
	8.3	Building Better Just Culture in a World with AI 132
		Lea-Sophie Vink
	8.4	Safety Management Systems Case Study 133
		Anastasios Plioutsias, Dimitrios Ziakkas,
		and Konstantinos Pechlivanis
		8.4.1	Case Study Overview 133
	Reference 134

Chapter 9 Best Practices for the Implementation of Artificial Intelligence 135

9.1 Common Challenges and Limitations of Implementing
AI in Transportation 135
Konstantinos Pechlivanis

References 135

9.2 Best Practices for Successful Implementation of AI in
Transportation, including Risk Assessment, Change
Management, and Collaboration 136
Lea-Sophie Vink

    9.2.1 Introduction 136
    9.2.2 Automation as the Concept of Operations 137
    9.2.3 The Levels of Automation 138
    9.2.4 Conclusions 138

References 138

9.3 Similarities and Differences between the Application of
AI in Automotive and Aviation Industries 138
Spyridon Chazapis and Dimitrios Ziakkas

References 140

9.4 Implementation of Requirements Systems – Start Now or
Miss the Chance 140
Anastasios Plioutsias and Dimitrios Ziakkas

Chapter 10 Future Directions and Emerging Trends in Artificial
Intelligence in Transportation 141

    10.1 Future Directions in AI Technologies 141
*Anastasios Plioutsias, Dimitrios
Ziakkas, and Konstantinos Pechlivanis*

Reference 141

    10.2 Emerging Trends in AI Technologies 141
*Anastasios Plioutsias, Dimitrios Ziakkas,
and Konstantinos Pechlivanis*

Reference 142

    10.3 Predictions and Prospects for the Future of AI in
Transportation 142
*Anastasios Plioutsias, Dimitrios Ziakkas, and
Aspasia Argyrou*

Reference 143

    10.4 Big Data and Safety Intelligence Possibilities 143
*Dimitrios Ziakkas, Anastasios Plioutsias,
and Dimitra Eirini Synodinou*

Reference 143

    10.5 The Advanced Air Mobility Case Study: Regulatory
Hurdles and Compliance Challenges 143
*Dimitrios Ziakkas, Anastasios Plioutsias,
and Yolanda Vaiopoulou*

    10.6 Conclusions 144
*Dimitrios Ziakkas, Anastasios Plioutsias, and
Konstantinos Pechlivanis*

Index 145

Preface

In the dynamic realm of transportation, the integration of artificial intelligence (AI) stands as a pivotal force reshaping not just the infrastructure and operations but also human performance and interaction within this sector.

Artificial Intelligence in Transportation: Applications, Challenges, and Future Directions delves into the intricate mesh of AI technologies and their profound impact on the way we understand, design, and engage with transportation systems. The emergence of autonomous vehicles, intelligent traffic management systems, and tailored mobility solutions urges us to thoroughly evaluate the opportunities and difficulties brought about by AI in the transportation sector. This book aims to illuminate the multifaceted implications of AI advancements, offering a comprehensive exploration of how these technologies enhance, complement, and sometimes challenge human roles and capabilities in the transportation domain.

The reader will go through a nuanced perspective on the AI-driven transformation in transportation, examining its effects on safety, efficiency, and sustainability, alongside the ethical, legal, and social considerations that accompany the adoption of such technologies. Our journey through the pages of this book is guided by contributions from leading experts across various fields including aviation science, engineering, computer science, urban planning, psychology, and ethics. Together, they offer valuable insights into the current state of AI in transportation, its prospects, and the intricate dance between human intuition and algorithmic precision. Through case studies, theoretical analyses, and empirical research, the chapters collectively paint a picture of a future where AI not only augments human performance but also fosters a deeper understanding of our relationship with technology.

Artificial Intelligence in Transportation: Applications, Challenges, and Future Directions is designed for academics, industry professionals, policymakers, and students who are navigating the complexities of this rapidly evolving landscape. It aims to serve as a foundational reference – a guide that not only informs but also inspires thoughtful consideration of how we can harness AI to create transportation systems that are safe, efficient, and in harmony with the human element. It is our hope that readers will gain a richer appreciation for the potential of AI in transforming transportation – a domain so integral to the fabric of modern society. This book is an invitation to engage with the questions, challenges, and possibilities that lie ahead and to contribute to shaping a future where technology and human performance coalesce to redefine the boundaries of what is possible in transportation.

Foreword

In 2023, generative artificial intelligence gained widespread recognition and popularity among the general public due to the remarkable achievements of ChatGPT, OpenAI's AI chatbot. There is undeniable evidence that a new era of industry has commenced.

The first industrial revolution, which concluded in the mid-19th century, marked a global shift from a labor-intensive economy to one that embraced more efficient and stable manufacturing methods. In contrast, the second industrial revolution, occurring from the late 19th century to the early 20th century, was characterized by a period of rapid scientific advancements, standardization, mass production, and industrialization. The significant proliferation of railway and telegraph networks following 1870 facilitated an unparalleled mobility of individuals, commodities, and concepts. Concurrently, novel technological infrastructures like electricity and telephones were deployed. The transition from mechanical and analog electrical systems to digital electronics, which began in the second half of the 20th century, including the introduction and general availability of digital computers and the digital management of documents, characterizes a third industrial revolution, which also continues today.

A new, fourth phase of major industrial changes, characterized by the convergence of technologies such as artificial intelligence, genetic engineering, and advanced robotics and with increasingly blurring boundaries between the physical, digital, and biological worlds, appears to be in full swing. As with previous revolutions, it is a revolution that will once again forever change the way people experience the world around them. Artificial intelligence is already being used, often without our noticing, in many aspects of our lives, such as more precise weather forecasts, early diagnosis of disease symptoms, or the selection of the most suitable candidates for a job, to name just a few.

As will become apparent when reading this book, the applications of AI are also particularly numerous in the broad domain of transport. Emerging technologies in AI, such as machine learning, natural language processing (NLP), and computer vision, offer the potential for exciting developments in the field of transportation. Machine learning systems can learn from data and optimize performance over time without the need for explicit programming. For example, machine learning algorithms can analyze historical data for maintenance and, combined with operational parameters, proactively predict maintenance, which can significantly increase the availability of assets. Or, based on the analysis of real-time transport data, the optimal routes and speeds for cars, trains, or planes can be predicted, and delays can be avoided. NLP allows computers to understand, interpret, and generate human language (written or spoken). Applications can actively support both users and operators, and through the analysis of safety reports, incident data, and maintenance logs, patterns can be recognized, and potential safety risks can be identified. Computer vision refers to the analysis and understanding of visual data by computers. If the saying "a picture is worth a thousand words" also applies here, the possibilities seem limitless, and a shift in the control of operational processes for safety and security from a reactive to a proactive and anticipatory approach can be anticipated.

The ever-accelerating introduction of AI therefore holds the promise of increasingly autonomous transport systems, in terms of sustainable performance and (personalized) user experience, more efficient maintenance, and improved safety. Within this last area, the use of big data also seems very promising. By harnessing the power of big data analytics, organizations can gain a deeper understanding of safety-related issues, predict, and prevent accidents, and continuously improve safety and overall performance, resulting in a more sustainable transportation industry.

Specifically, in the area of improved, sustainable, and safe performance of the railway system, a thorough introduction of AI offers opportunities in many areas: the maintenance of assets for both fixed installations and rolling stock will be better predicted; the route setting of train paths can be further automated; better monitoring of real-time activities allows for more proactive intervention in risky situations, especially for the monitoring of human performance; and so much more. What is also specific to railways is that it remains a challenge to find the right degree of harmonization among the various technical regional solutions that currently exist. Creating an evolution at different speeds is a risk around the corner, and increased attention will be needed to guide the transition.

However, promising the introduction of AI and transport may be, there are still quite a few uncertainties and points of attention, in railways as well as in other transportation modes. The current method of risk identification and safety certification of products and systems, which assumes an almost total knowledge of their functioning, clearly falls short of generative systems and will need to be rethought. Attention should also be paid to the role of human operators, who will continue to be needed to monitor the transport systems. Rather than a linear transformation from human tasks to automated technologies, new systems will have to be devised that strive for maximum resilience against unforeseen circumstances, based on the cooperation between man and machine that makes optimal use of the strengths of both components. Finally, attention should also be paid to the cultural and ethical aspects of introducing and, especially, accepting AI. We do not trust what we do not understand, and gaining the trust of operators in the coming years will be decisive for the immediate adoption of AI-driven changes. In addition, as AI continues to permeate the transportation industry, ethical considerations will become increasingly important. The industry will have to focus on ensuring fairness, transparency, and accountability in AI algorithms and decision-making processes. Efforts will need to be made to address potential biases, ensure data privacy and security, and establish regulations and standards to govern the responsible use of AI in transportation systems. Nevertheless, these concerns will not stop progress because there is a clear need (just think of the changing climate, an aging society, etc.) and the potential is enormous.

The forward march of technology will continue, and the challenge is to structure it and bring it to the benefit of society. This book provides a clear overview of AI's potential to change transportation industries for the better, from improving operational efficiency to revolutionizing the user experience to rethinking proactive safety management. The latter topic is an interest I share with the editors of this book, not only at the level of research orientation but most of all in the overall commitment to have safety at the center of organizational decision-taking. What makes this book unique is that it not only covers the potential of AI but also provides an honest view of the remaining challenges, with a clear insight into future trends and strategies for successful implementation. Moreover, the understandable and digestible way in which it is written also makes it accessible to all readers. By helping readers navigate what is possible and what is unfeasible or not possible with the current state of AI, the authors offer a series of stimulating discussion topics that can lead to concrete development and sustainable change. This book is therefore highly recommended, not only for upper management and decision-making levels but also for anyone who wants to become acquainted in a practical way with what AI has to offer the transport sector in the future.

Bart Accu, February 2024
Head of Support to Greece Task Force
European Union Agency for Railways

Clearly defining objectives is essential for effectively incorporating artificial intelligence (AI) into the transportation ecosystem. My research concentrates on implementing emerging technologies in transportation operations, particularly in areas such as safety, online training, autonomous operations, and economic strategies for software and hardware adaptation in civil military initiatives. An STS (Science and Technology Studies) approach emphasizes a comprehensive perspective on the problems of implementing AI in transportation. Over the past 2 years, I have concentrated on integrating aviation electronics and related technologies into single-pilot operations. This includes utilizing AI (VR-AR-MR-SATCE) digital twins for simulation and establishing certification and training standards for Advanced Air Mobility (AAM). While conducting research, I observed the fierce competition between Europe and the USA, as well as the emergence of new global actors in the aviation business following the Cold War. My primary purpose is to establish a worldwide network of entrepreneurs focused on implementing artificial intelligence in transportation. The network aims to improve safety and security, decrease operating expenses, and enhance the passenger experience. This book focuses on implementing the blue ocean operations model and tries to communicate this vision to all stakeholders through an ethical Corporate Social Responsibility (CSR) approach. It involves enlisting subject matter experts, regulators, and industry partners to participate in this transformation.

Dimitrios Ziakkas, West Lafayette, USA, February 2024

Artificial Intelligence in Transportation: Applications, Challenges, and Future Directions is a book that captures the essence of the paradigm shift in the transportation industry. Leading field experts author it and provide comprehensive insights into the multifaceted intersections of AI and transportation. This book covers various applications of AI in transportation and addresses its challenges. Moreover, it illuminates the exciting paths forward for the industry to take.

This book delves into the effects of AI on transportation. It covers diverse topics such as self-driving cars, aviation, trains, maritime, logistics, safety, and passenger experience. Additionally, it discusses the ethical, social, and regulatory aspects of AI deployment, stressing the importance of responsible use. This book aims to guide researchers, practitioners, and policymakers in navigating the rapidly evolving AI era of transportation.

Anastasios Plioutsias, Coventry, UK, February 2024

Acknowledgments

The editors gratefully acknowledge the assistance and contributions of the following organizations and people:

- Purdue University School for Aviation and Transportation Technology, for providing Assistant Professor Dr. Dimitrios Ziakkas the academic freedom to work on *Artificial Intelligence in Transportation: Applications, Challenges, and Future Directions*.
- Coventry University for providing Assistant Professor Dr. Anastasios Plioutsias the academic freedom to work on *Artificial Intelligence in Transportation: Applications, Challenges, and Future Directions*.
- Captain Konstantinos Pechlivanis and HF Horizons for their human-centric approach to emerging technologies.
- Dr. Lea-Sophie Vink for her academic and operational-minded contributions.
- Professor Dr. Michail Diakomihalis for his valuable contributions and comments.
- Editors and colleagues at the Publishing House for their guidance and responsiveness throughout the book development process.
- Contributors and transportation ecosystem SMEs who voluntarily supported this effort.
- Our families, for their words of encouragement and understanding despite the many hours we devoted to this work.

About the Editors

Dimitrios Ziakkas is an Assistant Professor in the School of Aviation and Transportation Technology at Purdue University. He is an Airbus captain, instructor, and examiner with current A-320-, A-330/350-, and A-340-type ratings. A cross-cultural communicator, global citizen, and innovator, Dimitrios has over 33 years of experience in military and civil aviation operations worldwide, integrating workforce planning, selection, and recruitment aviation practices in multi-disciplinary research programs. He retired from the Hellenic Air Force in 2008 and worked as a commercial pilot in private jets for Olympic Air, Qatar Airways, and Legend Airlines. As a pilot, instructor, and examiner, he has 12,000 total flying hours of experience with aircraft types such as the McDonnell Douglas F-4E Phantom II, Lockheed C-130 H/B Hercules, Gulfstream G-V, Premier 1/A (RA-390), Airbus A-320, A-330, A-340, and A-380.

During his service with the Hellenic Air Force, he was an electronic war officer and a flight safety investigator officer for 10 years. Olympic Air was head of the ICAO-level 4-language project. Qatar Airways participated in training and organized workforce planning, recruitment, and selection processes. He was Head of Training in Flight Training Organizations/Type Rate Training Organizations, Flight Instructor, Type Rate Instructor, Examiner, Crew Resource Management Instructor in Greece, and U.A.E. under JAR/EASA organizations. He earned a B.S. in Aviation Science in 1994 from the Hellenic Air Force Academy, Athens Greece, and a B.S. in Economics from the National and Kapodistrian University of Athens, Athens (NKUA), Greece, in 2003. Additionally, he earned an M.S. in History and Philosophy of Science and Technology (NKUA), Greece, in 2006. He holds a Ph.D. at the Graduate Program in the History and Philosophy of Science and Technology, NKUA, Greece (2014). Furthermore, he holds an M.S. in Human Factors in Aviation from Coventry University, UK (2016); an MBA from Birmingham University, UK (2019); and an MS in Human Resources Management and Development from Salford University, UK (2019).

He is a member of the European Union Aviation Safety Agency (EASA), the European Association for Aviation Psychology (EAAP), the Human Factors and Ergonomics Society (HFES), the Society for the History of Technology (SHOT), and a fellow member of the Royal Aeronautical Society, UK. His research is focused on the adoption of technology in the aviation operations area, flight safety and training, online flight training, single pilot operations, and the economical approach of software and hardware localization in the formation of aviation and military projects. He works on a project in the appropriation of aviation electronics, digital twins, and related technologies in single pilot operation: AI (VR-AR-MR-SATCE) simulation and formation of certification; CBTA (Competency-Based Training and Assessment) training requirements for Advanced Air Mobility.

About the Editors

Anastasios Plioutsias holds the position of Assistant Professor at the School of Future Transport Engineering, Coventry University, UK. He is an experienced pilot and has worked as an accident and incident investigator in military and civil aviation operations. His areas of expertise include Human Factors, Aeronautics, Aerospace Engineering, Ergonomics, Aviation Management, and Project Management Procedures. During his service with the Hellenic Air Force, he was a combat-ready fighter pilot accumulating more than 3000 hours of flying experience. Anastasios has flown several aircraft types, including the McDonnell Douglas F-4E Phantom II, Lockheed F-16BLK-52+, and various civil and general aviation types. He was also assigned as an accident and incident investigator and was part of all accident investigation and analysis boards. Additionally, he served as an intelligent officer. Since 2015, he upgraded and updated all the respective flight safety and ground schools of the Hellenic Air Force. He taught Crew Resource Management, Team Resource Management, Maintenance Resource Management, Safety Management Systems, Human Performance Military Aviation, and Accidents and Incident Investigations methods and tools. Furthermore, he was a flight and simulator instructor in the aircraft F-16C/D. He retired in 2018 to follow an academic career as an assistant professor of Safety and Human Factors at the Amsterdam University of Applied Sciences.

Anastasios obtained a B.Sc. in Aviation Science from the Hellenic Air Force Academy in Athens, Greece, in 1996. Later, in 2013, he earned a BSc in Business Administration and Management from the University of Thessaly in Larissa, Greece. In addition, he completed an MSc in Project Management and Procedures from the University of Thessaly, Larissa, in 2018. He recently obtained a PhD in Engineering and Ergonomics from the School of Mechanical Engineering at the National Technical University of Athens (NTUA) in Greece in 2021. He is affiliated with several organizations related to aviation safety and human factors, including the European Union Aviation Safety Agency (EASA), the European Association for Aviation Psychology (EAAP), the Human Factors and Ergonomics Society (HFES), Europe, the International Society of Air Safety Investigators (ISASI), and the Hellenic Ergonomics Society. He is a Fellow of the Royal Aeronautical Society (UK) and the Safety and Reliability Society (UK). His research is focused on adopting technology in aviation operations, flight safety and training, online flight training, and single-pilot operations. In recent years, he has collaborated with other universities and organizations in the UK, Europe, and the United States of America in areas such as Human Factors in Aviation, Artificial Intelligence and Human Factors (AI & HF), Unmanned Aircraft and Aviation Systems (RPAS, UAVs, and Drones), Connected Autonomous Vehicles (CAVs), the improvement of the design of new aircraft and systems, and investigations of aviation accidents and incidents. Currently, he is participating in similar research programs. He is a research associate of the University of Patras, the School of Mechanical Engineering, the Technical University of Crete, the School of Production Engineering and Management, and the NTUA School of Mechanical Engineering.

Contributors

Aspasia Argyrou
National and Kapodistrian University of Athens
Anthousa, Attiki, Greece

Ioannis Bakopoulos
Hellenic Air Force
Athens, Attiki, Greece

Volodymyr Bilotkach
Purdue University, School of Aviation and Transportation Technology
West Lafayette, Indiana

Thanasis Binis
National Aviation Investigation Agency and Railway Accidents and Transportation Safety (HARSIA)
Athens, Greece

Andrew Black
Coventry University
Eastleigh, Hampshire, UK

Spyridon Chazapis
HFHorizons Human Factors SME
Ano Glyfada, Athens, Greece

Caterina Ciacatani
Sixty Social Seconds
Dubai, United Arab Emirates

Stefano Conte
Coventry University
Forlì, FC, Italy

Evey Cormican
Visionary Training Resources (VTR) Inc.
Palm Harbor, Florida

Michail Diakomihalis
International Hellenic University & Hellenic Open University
Thessaloniki, Macedonia, Greece

Shantanu Gupta
Purdue University, School of Aviation and Transportation Technology
West Lafayette, Indiana

Debra Henneberry
Purdue University, School of Aviation and Transportation Technology
West Lafayette, Indiana

Konstantina Ioannidi
Faculty of Engineering and Architecture
Metropolitan College
Athens, Greece

Ioannis Katsidimas
YliSense PC
Patras, Greece

Julius Keller
Purdue University, School of Aviation and Transportation Technology
West Lafayette, Indiana

Anastasia Kosmidou
Human Factors Horizons
Bucharest, Romania

Athanasios Kotzakolios
YliSense PC
Patras, Greece

Shuo Liu
Northern Arizona University
Flagstaff, Arizona

Contributors

Natalia Marcovici
Bucharest Romania

Konstantinos Pechlivanis
Human Factors Horizons
Bucharest, Romania

Anastasios Plioutsias
Coventry University
Coventry, West Midlands, UK

Georges Rebender
Independent Expert
Koln, Germany

Timothy D. Ropp
Purdue University, School of Aviation and Transportation Technology
West Lafayette, Indiana

Konstantinos Sakalis
Department of History and Philosophy of Science
National and Kapodistrian University of Athens
Athens, Greece

Dimitra Eirini Synodinou
Human Factors Horizons
Polyguros, Halkidikis, Greece

Yolanda Vaiopoulou
Freelancer Nurse
Larissa, Thessaly, GR

Lea-Sophie Vink
Austro Control/Justminds AT
Vienna Austria

Beatrix Walzl
Vienna, Austria

Dimitrios Ziakkas
Purdue University, School of Aviation and Transportation Technology
West Lafayette, Indiana

Introduction

This book serves as a primary resource for addressing the practical aspects of implementing and ensuring the safety of innovative technologies in aviation and transportation. As of April 2023, worldwide searches for 'chat-GPT' and other AI tools like image generators and automatic software coding had surged to a record high. Every media outlet is abuzz with news about the upcoming wave of AI tools set to hit the market. Socially and culturally, we are grappling with philosophical questions related to overreach, data protection, and influence on thinking and social development. In various sectors, senior executives and authorities are consistently incorporating popular terms into discussions at events, investor gatherings, and client proposals to present a fresh perspective on transportation that addresses environmental worries and growing transportation needs and tackles challenges related to an aging and shrinking workforce.

The drafting of legislation has commenced. In 2021, the first draft of guidelines for the artificial intelligence concept paper was released by the European Union Aviation Safety Agency (EASA). It has been upgraded to a second version that is now available after undergoing an open review process. Given the current 'hot topics' being discussed as universal solutions and the delay in regulatory guidance, it is essential to provide practical advice to organizations as they navigate operational changes, address recruitment and training challenges, and implement AI tools for monitoring performance.

This book is designed to provide a dynamic and up-to-date solution with practical tips from experts in the field for upper management and decision-making levels. The European Association of Aviation Training and Educational Organizations (EATEO) and Purdue's School of Aviation and Transportation Technology (SATT) Department have recognized a research gap and the necessity for implementing AI in transportation. The following chapters highlight the most urgent issues that stakeholders in the transportation industry are facing. This guide will receive endorsements from several key organizations, such as the European Association for Aviation Psychology (EAAP). We intend to promote this book everywhere and gain endorsements from different organizations and institutions across the globe. The key to embracing AI lies in the culture and performance of individuals. Building trust with operators is crucial for ensuring the smooth adoption of technology in the future. It all begins with our approach to discussing technologies. Our goal with this book is to present topics in a clear and easy-to-understand manner that is accessible to everyone. We trust that the subjects will provide inspiration and discussion points for genuine progress and transformation, with a crucial emphasis on the necessary focus to distinguish between what is achievable and what is not.

Dimitrios' enthusiasm for integrating human-centric emerging technologies in aviation found the perfect platform at EATEO and Purdue's SATT to convene global leaders to participate. Anastasios, an ex-military pilot, technical expert, and professor at Coventry University, UK, is enthusiastic about studying human behavior, particularly in everyday situations. His extensive international background in content and context was instrumental in driving the project to completion on schedule and motivating the entire team.

Artificial Intelligence in Transportation: Applications, Challenges, and Future Directions will be the most comprehensive and concise book in its category. The ten chapters are organized into sections to enhance understanding of AI; implications for the transportation industry; contributions from science, technology, and society; change management and awareness; real-world AI implementation case studies; and future outlook. Reading each chapter individually is beneficial, but reading the book as a whole will provide a more comprehensive understanding.

1 Artificial Intelligence in Aviation

1.1 CONCEPTUAL FRAMEWORK FOR ARTIFICIAL INTELLIGENCE IN TRANSPORTATION

Konstantinos Sakalis

We live in smart homes equipped with intelligent doors and smart lights. We are surrounded by smart televisions, smart washing machines, and even toothbrushes with the so-called artificial intelligence (AI). Algorithms, which everyone claims they could learn, suggest movies and music. Our work involves smart assistants and intelligent schedulers reminding us of our appointments. Our communication is through smartphones and digital social media apps. We move using smart bicycles and commute to work in artificially intelligent cars guided by digital maps. At the same time, we face risks from smart weapons, autonomous drones, smart bombs, and dummy robots (Broussard, 2019).

The term 'Artificial Intelligence' was first used in 1956 at Dartmouth during a symposium focused on studying and creating 'intelligent' machines (McCarthy et al., 1955). Since then, its scope has significantly broadened. From earlier times to the present day, it encompasses a broad spectrum of hardware and software and a rich array of expectations, dreams, and aspirations. The multitude of definitions complicates rather than clarifies the concept of AI, blurring the boundaries of understanding and making historical semantic excavations somewhat challenging. Among them, recently, the European Union Aviation Safety Agency (EASA) has moved to a wider-spectrum definition that AI is a "technology that can, for a given set of human-defined objectives, generate outputs such as content, predictions, recommendations, or decisions influencing the environments they interact with (EASA, 2023)."

Long before the historical emergence of the concept of AI, societies, through various actors including scholars, policymakers, and technology evangelists, attempted to attribute 'intelligence' to objects. Whether it was through logarithmic rules (Tympas, 2017) or later through electronic circuits, personal computers, and at a more abstract level, algorithms, there was a pursuit to render 'intelligence.' Thus, preceding the prevalence of the concept of AI, terms such as 'intelligent robots,' 'automatic machines and automata,' 'explorers,' 'brain engineers,' 'electric brains,' and even 'steel brains' were encountered. In the field of transportation, a future with technologically advanced transportation options was presented to the American imagination as the 1939–1940 New York World's Fair promised to show visitors "the world of tomorrow." The Futurama at the General Motors Pavilion was a highly popular section of the exhibition. More than 24,000 visitors per day were drawn to

Futurama, eagerly waiting for hours to witness a glimpse of future transportation and hear the narrator's description of advanced transportation planning, featuring automated vehicles with radio controls navigating the city effortlessly.

As we know it now, the notion of AI can be traced back to Alan Turing's article 'Computing Machinery and Intelligence.' Turing (1950) inquires about the possibility and circumstances under which machines can exhibit thinking capabilities. He also presents the well-known Turing Test as a standard for addressing this issue. The history of AI begins with the first meeting at Dartmouth in 1956. John McCarthy and Marvin Minsky planned and secured financing for a two-month workshop that would involve leading scholars in the technology sector. However, the actual use of the term 'artificial intelligence,' as it is referenced concerning John McCarthy's selection, was, among other reasons, intended to avoid and detach from the prevailing term of 'Cybernetics' (McCarthy, 1996):

> As for myself, one of the reasons for inventing the term "artificial intelligence" was to escape association with "cybernetics". Its concentration on analog feedback seemed misguided, and I wished to avoid having either to accept Norbert (not Robert) Wiener as a guru or having to argue with him.

A depiction of the history of AI involves waves or periods of heightened attention and subsequent 'AI Winters.' As Garvey (2018) notes, the narration of AI history by its technology advocates leads to stories focusing on AI's benefits and successes, overlooking the risks and failures of systems to deliver their initial promises. The unfulfilled promises and empty threats prompted the three booms in AI publicity along with the intermediate AI Winters (Figure 1.1.1).

According to this chronology, the first publicity boom began in 1950 due to the competition between the USA and Russia, resulting in funded research. This phase lasted until the early 60s, marked by the failure to fulfill promises, leading to limited funding. The first AI Winter ended with a change in funding approach, focusing on the development of systems capable of specific tasks while highlighting Japan's technological advancement as a threat. This shift brought the second publicity boom with the development of expert systems. The unmet threats led to the collapse of this wave and the arrival of the second AI Winter in the 90s. The inception of the third wave of publicity is characterized by the dominance of machine learning (ML), accompanied by associated promises. The 90s concluded with numerous publications heralding the arrival of a 'Post-Industrial Society' as an 'Information Society,' where humanity's

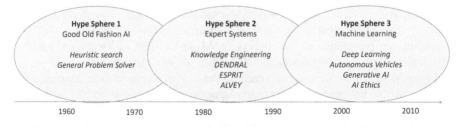

FIGURE 1.1.1   A timeline of AI's Hype Spheres.

daily problems would be solved with the aid of 'smart machines.' However, today, we find ourselves amid the discourse of witnessing the '4th Industrial Revolution,' where AI technologies emerge (once again) as a solution for every facet of our lives.

REFERENCES

Broussard, M. (2019). *Artificial Unintelligence: How Computers Misunderstand the World.* Cambridge: The MIT Press.

European Union Aviation Safety Agency (2023). *Artificial Intelligence Roadmap 2.0: Human-Centric Approach to AI in Aviation.* EASA. https://www.easa.europa.eu/en/document-library/general-publications/easa-artificial-intelligence-roadmap-20

Garvey, C. (2018). Broken Promises and Empty Threats: The Evolution of AI in the USA, 1956–1996. *Technology's Stories.* https://doi.org/10.15763/jou.ts.2018.03.16.02.

McCarthy, J., Minsky, M., Shannon, C. E., Rochester, N., & College, D. (1955). A Proposal for the Dartmouth Summer Research Project on Artificial Intelligence. http://jmc.stanford.edu/articles/dartmouth/dartmouth.pdf

Turing, A. M. (1950). I.-Computing Machinery and Intelligence. *Mind,* LIX(236), 433–460. https://doi.org/10.1093/mind/lix.236.433.

Tympas, A. (2017). *Calculation and Computation in the Pre-Electronic Era: The Mechanical and Electrical Ages.* London: Springer.

1.2   EVOLUTION OF AI AND ITS POTENTIAL APPLICATIONS
Konstantinos Sakalis

Since the emergence of the concept of AI in 1956, the accompanying rhetoric has spoken of societal transformation possibilities. Having traversed a period spanning over 60 years, its dominant (re)appearance in the public sphere underscores the demand for a critical evaluation of this trajectory, both on a conceptual and practical level, encompassing the realm of its artifacts. Following a conceptual chronology of the history of the electronic era, we can distinguish two distinct periods: the first encompasses the historical span from the early post-war decades until the 1980s, while the second period extends from the 1990s to the present day. This chronology is based on differences in materiality and the associated rhetoric regarding the 'capabilities' for social and regulatory possibilities of these technologies (Simos et al., 2023). During the first period (1950–1990), materiality was defined by the emergence of mainframes initially and personal microcomputers later. The relative rhetoric surrounding the presentation of these materialities with the term 'artificial intelligence' involved descriptors such as 'smart,' 'thinking,' and 'intelligent' (Natale & Ballatore, 2017). The second historical period (1990-present) is characterized by the widespread use of computers and their interconnected use, leading to the rise of the internet and social networks. While the use of descriptors like 'smart' and 'thinking' persisted during this period, there has been a generalized production of data, which is now presented as 'big data.' (Strasser & Edwards, 2017).

This chronology highlights the transition of AI from the capability of specific machines to the diffusion of intelligence across a network of interconnected

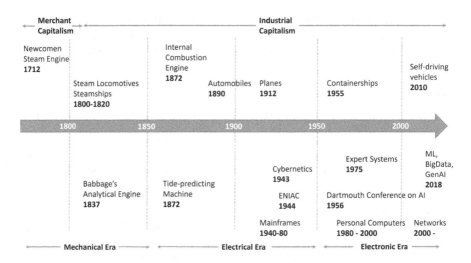

FIGURE 1.2.1 A parallel timeline of the evolution of transportation systems and artificial intelligence.

machines. Public discourse shifted from general issues, such as the use of computers at work, to ethical concerns, including the ethics of artificial intelligence (AI ethics). The public conversation returns to industrial revolutions through the lens of the Fourth Industrial Revolution (Figure 1.2.1).

In the landscape of transportation, the infusion of AI marks a pivotal juncture. The examination embarks on a historical journey through AI's evolution, unraveling its progression from rudimentary modes to the emergence of AI-powered mobility solutions. By scrutinizing significant milestones in AI technology, we could shed light on the profound global impact these innovations have had on societies and economies. Everyone can underscore the profound influence of AI in reshaping contemporary transportation, highlighting its potential to bolster safety, optimize operational efficiency, and fundamentally redefine the dynamics of mobility for both individuals and cargo. These findings represent a substantial stride in understanding the transformative prowess of AI within the transportation realm, steering us toward the horizon of safer, more efficient, and technologically sophisticated transportation systems poised to revolutionize the future of travel and logistics (Ziakkas et al., 2023).

Nowadays, transportation stands on the brink of a revolutionary transformation, poised to undergo an evolutionary leap with the integration of AI into its core systems. The convergence of advanced technologies and AI algorithms has ignited a change in basic assumptions in how we perceive, manage, and interact with transportation networks across the globe. As AI becomes increasingly intertwined with our vehicles, infrastructure, and logistics, it presents unprecedented opportunities to optimize efficiency, bolster safety, and revolutionize the entire transportation ecosystem. From the autonomous navigation of vehicles to predictive maintenance, traffic management, and logistics optimization, AI's applications in transportation are both diverse and expansive (U.S.D., 2023). The future of transportation is no longer a distant vision but a tangible reality taking shape before our eyes.

REFERENCES

Natale, S., & Ballatore, A. (2017). Imagining the Thinking Machine: Technological Myths and the Rise of Artificial Intelligence. *Convergence: The International Journal of Research into New Media Technologies*, 26(1), 3–18. https://doi.org/10.1177/1354856517715164.

Simos, M., Konstantis K., Sakalis K. & Tympas A. (2022). 'AI Can Be Analogous to Steam Power' or from the 'Postindustrial Society' to the 'Fourth Industrial Revolution': An Intellectual History of Artificial Intelligence. *ICON: Journal of the International Committee of the History of Technology*, 1, 97–116.

Strasser, B. J., & Edwards, P. N. (2017). Big Data Is the Answer . . . But What Is the Question? *Osiris*, 32(1), 328–345. https://doi.org/10.1086/694223.

U.S. Department of Transportation. (2023). *History of Intelligent Transportation Systems*. ITS Joint Program Office-HOIT 1200 New Jersey Avenue, SE Washington, DC 20590, www.its.dot.gov, History of ITS: FHWA-JPO-16-329.

Ziakkas, D., Vink, L.-S., Pechlivanis, K., & Flores, A. (2023). *Implementation Guide for Artificial Intelligence in Aviation: A Human-Centric Guide for Practitioners and Organizations*. ISBN 9798863704784.https://www.amazon.com/IMPLEMENTATION-GUIDE-ARTIFICIAL-INTELLIGENCE-AVIATION/dp/B0CL7Y4HPJ

1.3 BENEFITS OF AI IMPLEMENTATION IN TRANSPORTATION
Konstantinos Sakalis

AI is reshaping industries and streamlining operations with its wide-ranging applications across diverse sectors. In healthcare, AI aids in analyzing medical images, diagnosing diseases, and tailoring patient care. In finance, AI powers risk assessment models, algorithmic trading, and fraud detection systems. Moreover, generative AI (GenAI) enhances customer support through chatbots and virtual assistants, delivering efficient solutions and enhancing user satisfaction (IATA, 2023). Additionally, AI revolutionizes education with adaptive tutoring systems and personalized learning platforms, catering to the unique needs of each learner (Ziakkas et al., 2023). These examples underscore the versatility and impact of AI in driving innovation and efficiency across multiple domains.

1.3.1 AI ON TRANSFORMING TRANSPORTATION

The pioneering role of AI in transforming the transportation sector in several ways and making significant contributions could be found in the following key areas (IATA, 2018):

- First, AI is a crucial component in the development of self-driving or autonomous vehicles. ML algorithms enable these vehicles to perceive their surroundings, make decisions, and navigate safely without human intervention.

Furthermore, AI is used in driver assistance systems to enhance vehicle safety by providing features such as adaptive cruise control, lane departure warnings, and automatic emergency braking.
- Second, AI will affect the traffic management systems. AI helps optimize traffic flow by analyzing real-time data from sensors and cameras (Eurocontrol, 2020). Smart traffic lights can dynamically adjust signal timing based on traffic conditions, reducing congestion, and improving overall efficiency. Moreover, AI algorithms analyze historical traffic patterns and real-time data to predict future congestion and suggest alternative routes, optimizing travel times. In aviation, AI assists in air traffic control (ATC) systems by predicting optimal flight paths, managing air traffic congestion, and enhancing overall safety and efficiency.

AI is used to predict when transportation infrastructure, such as bridges, roads, and railways, requires maintenance. This proactive and predictive maintenance approach helps prevent unexpected failures, reduces downtime, and extends the lifespan of infrastructure (Garvey, 2018). Another field in which AI will be a transforming power is that of supply chain and logistics. AI optimizes route planning, fleet management, and delivery schedules, reducing fuel consumption, improving efficiency, and lowering operational costs. Predictive analytics helps in anticipating demand fluctuations, optimizing inventory levels, and streamlining the supply chain.

Furthermore, public transportation systems, electric and sustainable transportation, and ride-sharing services are affected by AI applications. AI is utilized in public transportation systems to improve scheduling, route planning, and the passenger experience. AI is used to optimize the charging infrastructure for electric vehicles by predicting demand and managing energy resources efficiently. Intelligent transportation systems help promote sustainable modes of transportation by encouraging the use of public transit, cycling, and walking. AI algorithms power matchmaking and optimization on ride-sharing platforms, improving the efficiency of matching drivers with passengers and optimizing routes for shared rides.

Finally, AI-based surveillance systems enhance security in transportation hubs, airports, and public spaces, helping detect and respond to potential threats. In addition, AI can analyze patterns to identify risky driving behaviors and enhance overall road safety (EASA, 2023). As AI technologies continue to evolve, their impact on the transportation sector is expected to grow, leading to more efficient, safe, and sustainable transportation systems.

REFERENCES

European Organisation for the Safety of Air Navigation (EUROCONTROL). (2020). The FLY AI Report Demystifying and Accelerating AI in Aviation/ATM. https://www.eurocontrol.int/publication/fly-ai-report

European Union Aviation Safety Agency. (2023). *Artificial Intelligence Roadmap 2.0: Human-Centric Approach to AI in Aviation*. EASA. https://www.easa.europa.eu/en/document-library/general-publications/easa-artificial-intelligence-roadmap-20

Garvey, C. (2018). Broken Promises and Empty Threats: The Evolution of AI in the USA, 1956–1996. *Technology's Stories*. https://doi.org/10.15763/jou.ts.2018.03.16.02.

Husbands P, Holland O & Wheeler M (eds.) (2008) The Mechanical Mind in History. Cambridge, MA, USA: MIT Press. http://mitpress.mit.edu/books/mechanical-mind-history

International Air Transport Association (IATA). (2018). AI in Aviation: Exploring the Fundamentals, Threats, and Opportunities of Artificial Intelligence (AI) in the Aviation Industry (White Paper). https://www.iata.org/contentassets/2d997082f3c84c7cba001f50 6edd2c2e/ai-white-paper.pdf

International Air Transport Association (IATA). (2023). Generative AI and Aviation. Report. https://www.iata.org/globalassets/iata/programs/innovation-hub/generative-ai-report. pdf

McCarthy, J. (1996). *Defending AI Research: A Collection of Essays and Reviews*. New York: Cambridge University Press.

McCarthy, John, Marvin Minsky, Claude Elwood Shannon, Nathaniel. Rochester, & Dartmouth College. (1955). A Proposal for the Dartmouth Summer Research Project on Artificial Intelligence. http://jmc.stanford.edu/articles/dartmouth/dartmouth.pdf

Natale, S., & Ballatore, A. (2017). Imagining the Thinking Machine: Technological Myths and the Rise of Artificial Intelligence. *Convergence: The International Journal of Research into New Media Technologies*, 26(1), 3–18. https://doi.org/ 10.1177/1354856517715164

Strasser, B. J., & Edwards, P. N. (2017). Big Data Is the Answer . . . But What Is the Question? *Osiris*, 32(1), 328–345. https://doi.org/10.1086/694223

Turing, A. M. (1950). I.-Computing Machinery and Intelligence. *Mind*, LIX(236), 433–460. https://doi.org/10.1093/mind/lix.236.433

Tympas, A. (2017). *Calculation and Computation in the Pre-Electronic Era: The Mechanical and Electrical Ages*. London: Springer.

Ziakkas, D., Vink, L.-S., Pechlivanis, K., & Flores, A. (2023). *Implementation Guide for Artificial Intelligence in Aviation: A Human-Centric Guide for Practitioners and Organizations*. ISBN 9798863704784

1.4 CONTROL ERGONOMICS IN THE ERA OF AI

Anastasios Plioutsias and Dimitrios Ziakkas

Control Ergonomics, also known as Human Factors Engineering, is the practice of designing systems that work well with their users' cognitive and physical abilities. Control systems have evolved since ancient times, progressing from fundamental mechanical levers and dials to intricate digital interfaces. With the advent of AI, these systems have undergone a revolutionary transformation by incorporating ML algorithms. The integration of ergonomics into control systems is essential for human-oriented design and efficiency in the era of AI (Tian, 2009; Tavana et al., 2016; Ponsa & Díaz, 2007). In today's AI era, ergonomics has expanded beyond just the physical interface to include cognitive interaction. Integrating AI technologies requires re-evaluating traditional ergonomic principles, emphasizing the need to adapt to the constantly evolving landscape of human-AI interaction.

The interactions between humans and AI systems are more complex than just using tools. AI systems are becoming more like collaborators, so humans must shift from direct operation to supervision and strategic decision-making. This change in approach is necessary because of the cognitive load on humans. Integrating AI into control systems can be beneficial, but it poses challenges. One such challenge is the possibility of cognitive overload and decision fatigue. The difficulties of balancing automation with human supervision emphasize the importance of designing systems that minimize these risks while maximizing the advantages of AI. Designing ergonomic AI systems involves following principles that guarantee user interfaces are easy to use, reduce mental effort, and enhance decision-making abilities (Ponsa & Díaz, 2007). Real-life examples and case studies demonstrate how such principles have been successfully applied in AI system design and unveil how such principles have improved efficiency, safety, and user satisfaction.

The potential ergonomic design developments utilize emerging technologies such as augmented reality and brain-computer interfaces. These advancements may further revolutionize the interaction between humans and AI and emphasize the importance of ongoing research and development. Integrating haptics into the overhead panel and control systems in aviation ergonomics enhances pilot interaction through tactile feedback, improving accuracy and safety. AI plays a significant role by dynamically adjusting these haptic responses based on flight data and pilot performance, optimizing the tactile guidance for various flight conditions. This symbiosis of haptics and AI not only streamlines cockpit operations but also significantly reduces the cognitive load on pilots, leading to a more intuitive and efficient control environment (Stanton and Baber 2006).

Integrating AI into control systems has brought a notable change in the field of ergonomics and shifted the focus from physical interfaces to cognitive interaction. This subchapter has explained the evolution, principles, challenges, and future directions of control ergonomics in the era of AI, emphasizing the importance of designing systems that improve human-machine collaboration. While AI technologies continue to advance, we need to update our approach to ergonomics to ensure we make the most out of these innovations while maintaining human well-being and efficiency.

REFERENCES

Ponsa, P., & Díaz, M. (2007). *Creation of an Ergonomic Guideline for Supervisory Control Interface Design* (pp. 137–146). https://doi.org/10.1007/978-3-540-73331-7_15.
Stanton, N., & Baber, C. (2006). The Ergonomics of Command and Control. *Ergonomics*, 49, 1131–1138. https://doi.org/10.1080/00140130600612523.
Tavana, M., Kazemi, M., Vafadarnikjoo, A., & Mobin, M. (2016). An Artificial Immune Algorithm for Ergonomic Product Classification Using Anthropometric Measurements. *Measurement*, 94, 621–629. https://doi.org/10.1016/J.MEASUREMENT.2016.09.007.
Tian, Y. (2009). Study on the Design of the Digital Control Machine-tool Operating System. *Mathematical Models and Methods in Applied Sciences*, 3, 101. https://doi.org/10.5539/MAS.V3N10P101.

1.5 LEARNING TO TRUST AI AND ML – KEY STEPS FOR IMPLEMENTING CHANGES

Dimitrios Ziakkas and Michail Diakomihalis

Learning to trust AI and ML in the context of transportation is an essential step towards implementing changes that can significantly enhance the efficiency, safety, and sustainability of transport systems. Trust in AI/ML technologies is fundamental to their acceptance and widespread use, from autonomous vehicles to traffic management and logistics optimization. The following table (Table 1.5.1) presents the proposed theories and provides frameworks for understanding how change can be effectively managed and how trust in innovative technologies can be cultivated.

TABLE 1.5.1
Implementing Change Theories

Theory	Description	Application to AI/ML Trust Building
Lewin's Change Management Model	Proposes change as a process with three steps: **Unfreeze, Change, and Refreeze.** It emphasizes preparing organizations for change, making the changes, and then solidifying those changes as the new norm.	**Unfreeze**: Educate stakeholders about AI/ML benefits and address fears. **Change**: Implement AI/ML technologies in pilot areas. **Refreeze**: Integrate AI/ML into standard practices and reinforce trust through consistent success stories.
Kotter's 8-Step Change Model	It focuses on eight steps to transform an organization, including creating a sense of urgency and forming strategic initiatives.	Apply by building a coalition for AI/ML adoption, creating a vision for its use, communicating the vision, and generating short-term wins to build momentum and trust.
ADKAR Model	A goal-oriented change management model that focuses on five outcomes: Awareness, Desire, Knowledge, Ability, and Reinforcement.	**Awareness**: Make stakeholders aware of the need for AI/ML. **Desire**: Foster a desire to embrace AI/ML technologies. **Knowledge/Ability**: Provide training on AI/ML. **Reinforcement**: Ensure ongoing support and celebrate successes to reinforce trust.
The Bridges Transition Model	Focuses on the transition process rather than the change itself, emphasizing emotional and psychological transitions.	Recognize the personal journey of accepting AI/ML in the workplace, providing support, and addressing concerns throughout the transition to build trust.
McKinsey 7-S Model	Outlines seven internal elements (Strategy, Structure, Systems, Shared values, Skills, Style, and Staff) that need to be aligned for successful change.	Ensuring all seven "S" elements support AI/ML integration can help build a cohesive environment where trust in AI/ML is part of the organization's culture.
Nudge Theory	Suggests positive reinforcement and indirect suggestions as ways to influence behavior and decision-making.	Use gentle prompts and success stories to nudge stakeholders towards trusting and accepting AI/ML technologies.

These theories provide many viewpoints and approaches for effectively handling the transition related to the adoption of AI/ML. The essential steps and areas of focus for addressing confidence in AI and ML, particularly in the transportation industry, are as follows:

Understanding AI and ML Capabilities: A foundational step in building trust is understanding what AI and ML can and cannot do. This involves demystifying AI technologies for stakeholders, including policymakers, transportation operators, and the public, by providing clear, accessible explanations of how these technologies work, their potential benefits, and their limitations.

The human brain can think in 15 minutes up to 15 ideas. But if the human mind is aided and abetted by an AI application, the number of these ideas jumps to 200 (Girotra et al., 2010). The Australian social researcher Mark McCrindle invented and used the term "Generation Alpha" to describe the new generation, born in the last decade of the 21st century. The characteristics of this generation are its technological intuitiveness, sense of global citizenship, and adaptability to change. (McCrindle, 2020). Generation Alpha is distinguished by an elevated level of familiarity and more comfort than previous generations, with AI, since members of this generation were born and grew up in an environment in which AI is part of people's daily lives and enters many aspects of modern society. Despite the rapid development of AI in recent years, companies remain hesitant and adopt its applications in various sectors with wariness, skepticism, and relative delay (Bughin et al., 2017). The systematic research on the subject will highlight further applications of technology in various functions of society and in many sectors of the economy.

The transport sector, but also that of shipping, are fields with significant scope for the application and use of AI, as well as in the related sectors that operate and serve these sectors, such as energy, manufacturing, wholesale commerce, etc. AI can be based on techniques with distinctive characteristics, such as the one that requires its operation to be fully supervised, partially supervised, and finally without any supervision (Brownlee, 2016). The handling and control of these techniques can be delegated to a series of algorithms that will result in the appropriate behavior and desired results.

The human factor supervises the process and depending on the results and the reliability of the results, the algorithm can be fed back, to adapt accordingly and improve the quality of the answers, after correspondingly adjusting its calculations in the future. Brownlee has further classified supervised learning as a "regression" problem to search for specific answers, while a simple "classification" simply classifies information into various categories.

Supervised learning techniques need, according to Castle (2017), a reliable set of data with the appropriate labeling from which the proper answers seek to come, and consequently, the capabilities of these data will be further improved and become more effective. Finding a solution to an issue in the traditional way and without the use of AI assistance will not be able to be communicated to specific users, since it is not possible to communicate this knowledge to all ships of the shipping company. The sought-after solutions should be available for algorithm feedback, and recognizable to the users looking for them. AI that is asked to process a large amount of complex information must be able to indicate to the user the origin of the data, so that it can assess the reliability of the data and its sources of origin and feed the algorithm accordingly.

The methodology for determining the first specific application areas of AI can have a similar approach and application as that already used in energy saving issues. That is, the gradual application of AI is proposed to start with easier and more controllable subjects and with projects of limited duration and predictable success (Capehart et al., 2016, pp. 23–24). The difficulty and the expected reaction from an application of AI are parameters that should be considered and if it is contemplated appropriate to have a modification in the way and the final form of application (Kotter and Schlesinger, 2008).

After the initial installation and operation of intelligent learning, with the basic and easier-to-understand applications, such as classification, archiving, easy retrieval, and access, it will be easier and more acceptable to further exploit AI for other more complex applications. According to the global AI research by McKinsey, it is revealed that the adoption of AI is constantly increasing with continuous technological returns (McKinsey & Company, 2023). The findings of the McKinsey survey show that there has been an annual increase of around 25% in the use of AI for purely business processes, with a parallel jump compared to the previous year for all enterprises already using applications of AI in various business process sectors. This study concluded that early adopters of AI, mainly by implementing a proactive strategy in the transportation and logistics sector, saw higher profit margins.

The impact of AI is not the same for all the activities it is applied to. For example, the logistics industry works with AI applications. Correspondingly, shipping activity also has immense potential for improvement by achieving better quality and speed in the operational activities of businesses. The emerging question posed by those who choose to implement and adopt AI focuses on how it can contribute to the further development of the shipping business and how it will lead the shipping industry to the next level of development. The self-evident answer to the above question concerns the significant opportunity provided to all those involved in maritime transport to further improve business operations with the application of AI, which can furnish greater accuracy in predicting the time of arrival and departure of the ship and in the more accurate estimation of fuel consumption, or even the possibility of saving fuel. Therefore, further development in the shipping industry, with the application of AI will contribute to the use of big data and the operation of ML, in which ship workers and business management can be trained. The fundamental operations of the administration of vessels on land and at sea (Personnel and crew management, Technical management, Commercial management, Bunkering, Dry-docking, and Financial management and accounting) are encompassed within the subsequent classifications in each of which AI and ML have the potential to enhance performance and effectiveness.

Demonstrating Safety and Reliability: For transportation, where safety is paramount, demonstrating the safety and reliability of AI systems through rigorous testing and real-world trials is essential. This includes simulating various operational scenarios, conducting pilot projects in controlled environments, and gradually introducing AI systems into public spaces while closely monitoring their performance.

Establishing Regulatory Frameworks: Developing and implementing comprehensive regulatory frameworks that address safety, privacy, accountability, and ethical considerations associated with AI in transportation can help build public trust. These frameworks should ensure that AI systems meet stringent safety standards, respect user privacy, and are designed and operated transparently.

Promoting Transparency and Explainability: AI systems should be designed to be as transparent and explainable as possible. Stakeholders should have access to information about how decisions are made by AI systems, particularly in critical situations. This transparency helps build trust by making AI operations understandable to humans and ensuring that AI systems can be audited and held accountable.

Engaging with Stakeholders: Continuous engagement with all stakeholders, including the public, industry experts, policymakers, and ethicists, is crucial for understanding concerns and expectations regarding AI in transportation. This engagement can take the form of public consultations, workshops, and forums that facilitate open discussions about the benefits and challenges of AI technologies.

Investing in Education and Training: Educating the workforce and the public about AI and ML technologies can demystify these systems and reduce apprehension. Training programs for transportation professionals can help them understand how to interact with and oversee AI systems effectively, ensuring these technologies are used responsibly and ethically.

Implementing Incremental Changes: Gradually implementing AI technologies in transportation, starting with less critical applications before moving to more sensitive areas, allows for the monitoring of performance and the building of trust over time. This stepwise approach enables adjustments and improvements to be made based on real-world experience.

In conclusion, building trust in AI and ML technologies in transportation requires a multi-faceted approach that includes education, transparency, rigorous testing, regulatory oversight, stakeholder engagement, and a gradual implementation strategy. By identifying the key steps/areas of change, the transportation sector can harness the potential of AI and ML to improve operations, enhance safety, and reduce environmental impact, all while maintaining public trust and acceptance.

REFERENCES

Brownlee, J. (2016). Supervised and Unsupervised Machine Learning Algorithms. *Machine Learning Mastery*. Application of Artificial Intelligence in the Maritime Industry. Retrieved from https://machinelearningmastery.com/supervised-and-unsupervised-machine-learning-algorithms

Bughin, J., Hazan, E., Manyika, J., & Woetzel, J. (2017). Artificial Intelligence: The Next Digital Frontier? Mckinsey Global Institute. Retrieved from https://www.mckinsey.com/~/media/McKinsey/Industries/Advanced%20Electronics/Our%20Insights/How%20artificial%20intelligence%20can%20deliver%20real%20value%20to%20companies/MGI-Artificial-Intelligence-Discussionpaper.ashx

Capehart, B. L., Turner, W. C., & Kennedy, W. J. (2016). *Guide to Energy Management*. Lilburn, GA: The Fairmont Press, Inc.

Castle, N. (2017.). Supervised vs. Unsupervised Machine Learning. Retrieved from https://www.datascience.com/blog/supervised-and-unsupervised-machine-learning-algorithms

Girotra, K., Terwiesch, C., & Ulrich, K. T. (2010). Idea Generation and the Quality of the Best Id-ea. *Management Science*, 56(4), 591–605. https://www.jstor.org/stable/27784139

Kotter, J., & Schlesinger, L. (2008). Choosing Strategies for Change. *Harvard Business Review*, 86(7,8), 130–139.

McCrindle, Mark. (2020). *Understanding Generation Alpha*. Norwest NSW: McCrindle Research Pty Ltd.

McKinsey & Company. (2023). *The State of AI in 2023: Generative AI's Breakout Year, Survey*. New York: McKinsey & Company.

1.6 AI CERTIFICATION & CHALLENGES
Georges Rebender

The aviation sector is currently utilizing AI in different identified areas. One domain area (case study) is the monitoring of the continued airworthiness of airplanes. A contemporary airplane is outfitted with several sensors that offer high-quality data. This data is examined by AI maintenance software programs utilizing AI ML/ supervised learning, which has been educated on past flight history.

Extending the scope of discussions to public acceptance, AI, like many innovations, raises many concerns. During the 19th century, the railway was introduced across Europe. Concerns were voiced by the public when railways were introduced, with some scientists predicting that speeds faster than twice the speed of a galloping horse could cause major heart problems for passengers. However, these fears were proven to be overblown based on experience. The advancement of AI tailored for a particular demographic is raising concerns, particularly over the risk of malevolent superintelligence, typically inspired by science fiction. Several prominent individuals have warned about substantial risks that could threaten the survival of humanity. Furthermore, in the realm of general aviation, where numerous aircraft are approved for single-pilot operation, the Garmin Autoland system, certified by both the FAA and EASA, can take over and safely land the aircraft in the event of pilot incapacitation.

1.6.1 WHY SHOULD WE REGULATE?

Regulators must protect society, including individuals and property, which is often referred to as the concept of precaution due to their awareness of public perception. Rules are necessary in all countries to safeguard residents from hazards or risks. The road example is intriguing: the driver may cause harm to himself or others inside or outside the car due to the energy of the vehicle. Thus, to minimize the risk to the lowest feasible level, possessing a driving license is a crucial aspect of risk management and is subject to licensing. Conversely, a bicyclist, while required to follow road traffic laws, typically does not need a license. The equation shows that reducing the energy of the bicycle will decrease the risk to the driver and others. The principle of proportionality, likely influenced by notions of freedom, also has an impact. To alleviate road traffic issues, one approach could involve implementing a risk-based licensing system for all modes of transportation. The aviation regulatory overarching concept embraces:

- Human-centered safety and security risk reduction as low as reasonably practicable.
- Principle of proportionality: aviation regulations consider factors such as aircraft weight, number of passengers, type of operations, public air transport or general aviation. In one word, one size does not fit all.
- Environmentally friendly conditions

1.6.2 AI REGULATORY APPROACHES

AI's scope is very wide as it covers numerous activities in different sectors that may have their specific particularities. AI can be used in transportation, management of infrastructure, health care, recruitment or selection of natural workers, biometric identification, financial services, education and vocational training, child safety, and law enforcement (IATA, 2018).

Various regulatory approaches can be considered: ***decentralized, centralized, or blended***.

Decentralized Approach: A decentralized approach is a sectorial bottom-up approach covering the specificities of each sector. The United States is actively shaping its approach to regulating AI, particularly within the aviation sector and beyond, through a series of executive orders, federal initiatives, and regulatory guidelines. The Biden-Harris Administration has taken significant steps to harness and manage the potential risks associated with AI technologies. Three months ago (November 2023), President Biden issued an executive order aimed at ensuring America's leadership in AI by directing actions to strengthen AI safety and security, protect privacy, promote innovation, and more. This initiative has led to substantial progress in mitigating AI risks and fostering innovation. The Federal Trade Commission has emerged as a key player in potentially shaping federal AI regulations. The Federal Trade Commission has issued publications indicating a growing focus on AI, emphasizing the importance of using representative data sets, testing AI for bias, ensuring explainability, and creating accountability mechanisms for AI development and use. These steps signal a broader regulatory interest in addressing the challenges posed by AI technologies.

Furthermore, the U.S. oversight of AI goes beyond federal efforts to encompass activities at the state level and the development of frameworks such as the AI Risk Management Framework by the National Institute of Standards and Technology (NIST). This framework serves as a voluntary guide for organizations to manage risks associated with AI and promote trustworthy AI systems, potentially setting industry standards. At the state level, legislation and regulations are also evolving to address AI applications, with states introducing bills and forming task forces to examine AI-centric regulations. The initiatives encompass several concerns such as facial recognition technologies, consumer rights, employment, automation in the finance and insurance business, and healthcare applications.

The changing U.S. regulation of AI demonstrates a sophisticated strategy that weighs the importance of reducing dangers against the goal of encouraging innovation and preserving America's dominance in AI technologies. This involves prioritizing AI safety and security, dealing with privacy issues, and creating a supportive environment for the ethical advancement and implementation of AI systems. As AI

Artificial Intelligence in Aviation

continues to transform industries, including aviation, the U.S. regulatory landscape is likely to keep evolving in response to new challenges and opportunities presented by these technologies.

Centralized Approach: This approach is a centralized horizontal top-down approach addressing all sectors. Therefore, it can only be a hard law, applicable to all domains included in the law. Europe has adopted this approach, and the outcome is now determined. The European AI Act was adopted on February 2, 2024, through the EU co-decision process, including the European Parliament and the European Council of Ministers. Based on a draft initial proposal from the EU Commission, the regulator considered the following options:

- Option 1: EU legislative instrument establishing a voluntary labeling program.
- Option 2: sector-specific ad hoc approach.
- Option 3: involves a horizontal EU legislative instrument that adopts a proportionate risk-based approach.
- Option 3+: involves the creation of a horizontal EU legislative instrument that adopts a proportionate risk-based approach and includes norms of behavior for non-high-risk AI systems.
- Option 4 proposes a horizontal EU legal instrument that sets necessary rules for all AI systems, regardless of the hazards they present.

Following co-decision debates, Option 3+ was chosen.

1.6.2.1 EU AI Act: High-Level Principles

Word of caution: The analysis is derived from the preliminary AI Act from April 2021. The AI Act finalized on February 2, 2024, has not been released to the public yet. Although the material is anticipated to remain the same, certain parts may undergo alterations.

Applicability:

- AI system providers are located in the EU or a third country that introduces AI systems to the EU market or uses them within the EU.
- EU-based users of AI systems.
- AI system providers and users in a non-EU country whose output is utilized within the EU.
- Users are responsible for following the producer's operating instructions, which is akin to aircraft operators' responsibility to adhere to aircraft manufacturer paperwork.

Finally, note that the majority of the AI Act is devoted to AI manufacturers.

Legal framework:

Concerns include risk-based safety, security, and health, as well as ethical considerations related to AI and compliance with EU data protection regulations. The risk-based approach is important to the EU legislation, and it will be further elaborated in 1.6.3.4.

Enforcement measures:

EU Member States must appoint a competent authority to oversee the enforcement of the rule and create an EU AI board consisting of members from EU Member States. Administrative penalties might reach up to 30 million euros or 6% of the entire global annual turnover.

Innovation support:

Provision rules create a regulatory sandbox to help construct, test, and validate AI systems before they are introduced to the market.

Dates of Entry into Force and Application:

The application typically takes place 24 months after going into effect, and 12 days following publication.

Grandfather rights:

AI systems now available for purchase or use: This law applies to high-risk systems that were introduced to the market or put into operation before the application date, only if they undergo substantial modifications in their design or intended use from that date onward.

1.6.2.2 Heart of the EU System: The Risk-Based Approach

The AI Act introduces a comprehensive risk classification system for AI applications, structured as a four-stage pyramid, addressing not only safety and security concerns but also ethical risks. These ethical risks include social scoring, the use of biometric identification systems (with exceptions for severe crimes), sensitive markers, predictive policing, and emotion recognition systems. The Act categorizes AI systems into four levels of risk: ***Unacceptable, High, Medium, and Low*** (Figure 1.6.1).

- *Unacceptable risks* are strictly prohibited due to their ethical implications, such as social scoring and non-essential use of biometric identification.

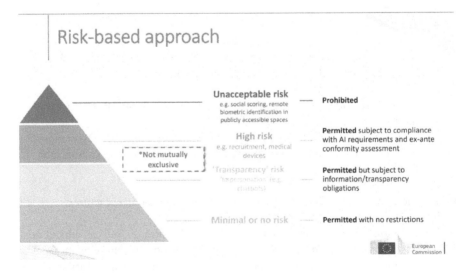

FIGURE 1.6.1 AI Act risk classification levels presentation.

Artificial Intelligence in Aviation

- **High-risk** systems are those related to safety or possessing ethical characteristics significant enough to warrant stringent regulation. These include AI applications in biometric identification, education, employment, essential services, law enforcement, border control, and justice administration. For such high-risk systems, the AI Act mandates a conformity demonstration, encompassing both ex-ante requirements (like risk management, data governance, and human oversight) and ex-post requirements (including post-market monitoring aligned with existing regulations like EASA for aviation).
- **Medium-risk** systems are subject to information and transparency obligations, ensuring users are aware of the AI's functionalities and limitations.
- **Low-risk** systems, deemed to pose minimal ethical or safety concerns, are not subject to regulatory obligations. This tiered approach allows for balanced regulation that promotes innovation while safeguarding ethical standards and public safety.

1.6.2.3 Technical Standards

Aviation safety regulations may not cover all the subtle details associated with AI because of the topic's complexity. High-risk AI systems require advancements in technical standards. The European Commission's actions involve the establishment of high-risk systems with standardized regulations, as seen in Figure 1.6.2.

1.6.3 EU AVIATION AI BLENDED APPROACH

This aviation strategy is a modification of the EU centralized approach. Although the centralized law upholds its principles and spirit, it does not directly apply to certain sectors such as aviation and cars. However, essential requirements for high-risk AI should be considered in relevant implementing or delegated acts issued by the EU Commission (AI Act explanatory memorandum). The need for harmonized international standards addressing safety and security such as SAE /EUROCAE WG114/ G34 or RTCA/EUROCAE WG72/SC216 will be discussed in **Chapter 9.1** AI challenges and limitations. The EC has been assigned by the legislature to create regulations for the implementation of aviation AI that comply with the requirements for

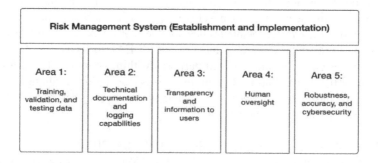

FIGURE 1.6.2 Risk Management Systems presentation.

high-risk systems outlined in Title III, Chapter 2 of the AI Act, relating to EASA fundamental regulation operations as detailed in specific articles. In details:

- EASA BR article 17/19 addresses implementing rules for **aircraft airworthiness.**
- EASA BR article 43/47 addresses implementing rules for **Air traffic management /air navigation systems.**
- EASA BR Article 57/58 addresses implementing rules for **unmanned aircraft.**

Note: EASA BR Article 88 addresses the general interdependencies between civil aviation and security.

EASA Case Study

ANTICIPATED REGULATORY STRUCTURE FOR AI

EASA's anticipated regulatory structure for AI management follows a top-down approach starting from the EU AI Act down to the industry standards (Eurocontrol, 2020). The first level reflects sectorial-level articles for which domain-specific implementing rules, such as aircraft airworthiness, drones, and ATM, as noted above, will be produced by the EU Commission if the published final act remains unchanged. Implementing rules can be accomplished through accepted ways of compliance, guidance materials, and references to agreed-upon industry standards. Other sectors, operations, flight crew licenses, and airports, except for airport security, may directly refer to industry standards under EU regulation 300/2008.

EASA PRELIMINARY AI AVIATION DEVELOPMENTS

In Europe, as a first step, the EU high-level expert group for AI established guidelines for trustworthy AI, which were supplemented by the Assessment List for Trustworthy Artificial Intelligence (ALTAI) for self-assessment. Through Innovation Partnership Contracts (IPCs), EASA was able to cooperate on AI initiatives. IPCs are a pre-application service provided by EASA to external stakeholders who wish to get technical advice before or outside of the certification procedure, in compliance with Commission Regulation 748/2012. Services include assisting with the implementation of a disruptive technology or novel concept whose feasibility may need to be proven and for which an acceptable framework does not yet exist or is not developed.

EASA has initiated a research project called MLEAP to focus on ML approval for systems designed for safety-related purposes within the scope of EASA's basic regulation, or the digital Europe 3 program. This is essential for enabling small and medium-sized businesses to access AI technology. The agency described its first AI experience in two advisory publications.

- EASA Artificial Intelligence Roadmap 2.0 (EASA, 2023).
- EASA Concept Paper Guidance for Level 1&2 Machine Learning Applications.

Artificial Intelligence in Aviation

EASA AI Taxonomy

EASA's technological scope is broad, encompassing not only ML and deep learning, but also hybrid AI, expert systems, and Bayesian estimation. This term also includes logic and knowledge-based (LKB) systems. Figures 1.6.3 and 1.6.4 present the EASA approach regarding taxonomy and the human-centered trustworthy approach (EASA, 2023).

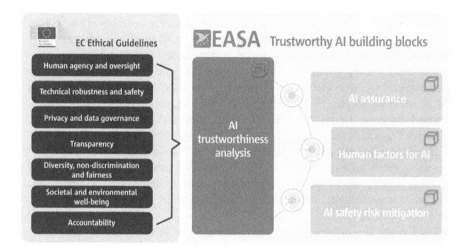

FIGURE 1.6.3 EASA trustworthy building blocks.

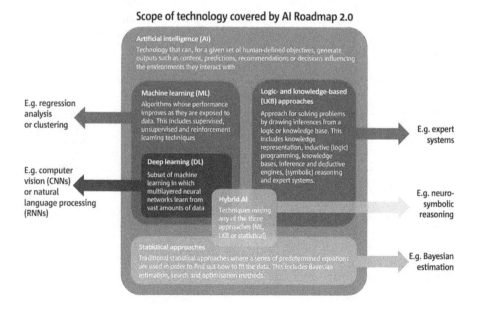

FIGURE 1.6.4 Presentation of AI taxonomy on the EASA AI roadmap 2.0.

Based on the EC ethical guidelines published by the EC high-level group, the trustworthy AI analysis addresses three main building blocks:

AI assurance: As with classical software, the AI application must demonstrate that it "safely performs its intended function" while being tailored to the AI statistical software features. As a result of its statistical character, the fundamental concern of AI is whether we can trust the output of an AI system. Or differently, what are the chances that the output is correct? The key elements are safety and security assessments and ethics-based assessments.

Learning assurance: Current development assurance approaches are not suitable for AI ML. Learning assurance addresses the move from programming to learning in this paradigm. The EASA concept paper suggests the W shape concept for guiding level 1 and 2 ML applications, as seen in Figure 1.6.5. The W-shaped concept, already used successfully by industry aviation experts such as human factors SMEs, has been adapted by EASA to AI ML training, validation, and test data. This approach may merit consideration in working groups discussing AI ML technical standards (Ziakkas et al., 2023).

Human factors for AI: Article 14 of the AI Act mandates human oversight of the human-machine interface to ensure that natural humans can effectively supervise the AI system during its operation. Human oversight of high-risk AI systems should prioritize the prevention or reduction of dangers in their intended usage or in situations of expected misuse.

The EASA AI proposed human-cantered risk classification follows three levels:

- Level 1 assistance to humans is probably the one that, in the short term, will attract most AI applications. The question of aviation AI application safety risk assessment regarding the AI Act risk classification is probably the most challenging issue to be tackled by the industry and regulators. Not all AI applications are high-risk-related, e.g., non-mandatory/non-essential AI applications related to optimization.

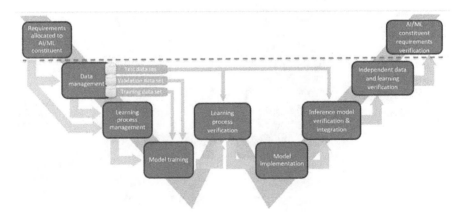

FIGURE 1.6.5 EASA human-centered trustworthy AI approach.

- Level 2 will introduce an operations and training human/AI paradigm shift worth developing in flight operations and in training by AI producers through a training area of special emphasis.
- Level 3, aiming for autonomy, raises the AI fundamental competency for autonomous AI situational and decision-making.

As the EU drone operations implementing rules allow autonomous operations, the first applications may be in the domain.

1.6.4 Conclusions

Aviation air transport is extensively regulated by ICAO/Chicago convention signatories. AI risk assessment could benefit from the extensive knowledge gained in aviation with the implementation of 1309 regulations (IATA, 2023). The current aviation software regulations in Europe are not explicitly outlined in high-level laws such as the EASA basic regulation or IR Part 21. Instead, they are referenced in EU Certifications 25, which are similar to FAR 25 paragraphs 1301d and 1309. These regulations focus on ensuring that the system functions as intended and consider failure consequences and probabilities of occurrence. DO-178C, "Software Considerations in Airborne Systems and Equipment Certification," certifies commercial software-based aircraft systems. Developed by RTCA and EUROCAE, it emphasizes ensuring software reliability and safety. The standard categorizes software according to risk levels, from unacceptable to high, medium, and low, detailing specific requirements for compliance and post-market monitoring. It updates the previous DO-178B version, addressing modern software development practices and technologies. When developing AI safety standards, it is important to prioritize pragmatism, common sense, proportionality, human-centeredness, trustworthiness, and adherence to internationally agreed-upon guidelines.

REFERENCES

European Organisation for the Safety of Air Navigation (EUROCONTROL). (2020). The FLY AI Report Demystifying and Accelerating AI in Aviation/ATM. https://www.eurocontrol.int/publication/fly-ai-report

European Union Aviation Safety Agency. (2023). Artificial Intelligence Roadmap 2.0: human-centric approach to AI in aviation. EASA. https://www.easa.europa.eu/en/document-library/general-publications/easa-artificial-intelligence-roadmap-20

International Air Transport Association (IATA). (2018). AI in Aviation: Exploring the Fundamentals, Threats, and Opportunities of Artificial Intelligence (AI) in the Aviation Industry (White Paper). https://www.iata.org/contentassets/2d997082f3c84c7cba001f506edd2c2e/ai-white-paper.pdf

International Air Transport Association. (IATA). (2023). Generative AI and Aviation. Report. https://www.iata.org/globalassets/iata/programs/innovation-hub/generative-ai-report.pdf

Ziakkas, D., Vink, L.-S., Pechlivanis, K., & Flores, A. (2023). *Implementation Guide for Artificial Intelligence in Aviation: A Human-Centric Guide for Practitioners and Organizations.* ISBN 9798863704784.https://www.amazon.com/IMPLEMENTATION-GUIDE-ARTIFICIAL-INTELLIGENCE-AVIATION/dp/B0CL7Y4HPJ

1.7 SIMULATED ATC ENVIRONMENT (SATCE) AND ASTi SERA's ROLE IN CBTA: THE PURDUE UNIVERSITY CASE STUDY

Dimitrios Ziakkas

Introduction to CBTA in Aviation Training

Competency-Based Training and Assessment (CBTA) represents a shift in aviation training towards measuring trainee performance based on specific competencies, rather than merely time spent on training tasks. This approach focuses on ensuring pilots acquire and demonstrate the necessary skills and behaviors for safe and efficient flight operations.

SATCE and ASTi SERA's Role in CBTA

The Simulated Air Traffic Control Environment (SATCE), particularly through the implementation of Advanced Simulation Technology Inc. (ASTi's) Simulated Environment for Realistic ATC (SERA), plays a pivotal role in CBTA by providing realistic ATC interactions. This technology enhances the training environment, allowing trainees to develop and demonstrate competencies in communication, situational awareness, and decision-making in scenarios connected with evidence-based training.

Purdue University's Case Study Objectives

The Purdue-ASTi research case study on SATCE offers a realistic and immersive environment for aviation SMEs to enhance their knowledge and skills. The Purdue University School of Aviation and Transportation Technology (SATT) team aims to study SATCE training scenarios and implement a CBTA approach in the communication competency (Ziakkas et al., 2023).

Implementation and Methodology

To achieve these objectives, Purdue University integrated ASTi's SERA technology into the A-320 MPS FSTD. This integration was designed to simulate realistic ATC communications and environments, thereby providing a platform for students to practice and demonstrate competencies in a controlled but realistic setting.

Findings and Impact on CBTA

The case study demonstrated that integrating SATCE - ASTi SERA significantly enhanced the CBTA process by providing students with opportunities to develop critical competencies in a realistic and immersive environment. The implementation of ASTi -SERA showed marked improvements in areas such as communication skills, situational awareness, and decision-making abilities.

Conclusion and Future Directions

The integration of SATCE - ASTi SERA into CBTA at Purdue University represents a significant advancement in aviation research and training. By providing a realistic and immersive training environment, this technology supports the effective development and assessment of pilot competencies, aligning with the goals of CBTA. Looking forward, the success of this case study suggests a promising future for the broader adoption of such technologies in competency-based aviation training programs, potentially setting new standards for pilot training and assessment.

REFERENCE

Ziakkas, D., Waterman, N., & Flores, A. (2024). *Emerging Technologies in Aviation: The Simulated Air Traffic Control Environment (SATCE) Application in Competency-Based Training and Assessment.* AHFE Open Access. https://openaccess.cms-conferences.org/publications/book/978-1-958651-95-7/article/978-1-958651-95-7_64

2 Artificial Intelligence Training and Operations

2.1 INCORPORATING AI INTO TRANSPORTATION HIGHER EDUCATION: CURRICULUM CONSIDERATIONS FOR AN EVOLVING INDUSTRY

Debra Henneberry and Julius Keller

Higher education institutions have a critical role in shaping the future of transportation. They achieve this by educating the upcoming generation of transportation professionals, conducting innovative research, and collaborating with industry stakeholders to decipher complex challenges and promote innovation. Universities are equipping students with the necessary skills to understand and address the challenges of the global transportation system. This is achieved through the adoption of interdisciplinary approaches, the incorporation of emerging technologies, and the provision of experiential learning opportunities. The ultimate goal is to develop sustainable, efficient, and equitable transportation networks for the future.

The incorporation of rapidly evolving technologies such as AI, data analytics, and autonomous systems has additionally revolutionized transportation education. Some universities are creating specific research centers, laboratories, and degree programs focused on exploring the implementation of AI in transportation, augmenting traffic efficiency, enhancing safety, and advancing autonomous vehicle technology. However, not all universities have the same resources, and some will be challenged to keep up with the speed of change. Some disciplines within transportation higher education are making changes to curricula faster than others. For example, engineering programs continue to update their curricula to include courses that engage students in smart mobility, data acquisition and analytics, autonomous driving, and logistics. Many of these programs adapt AI concepts through a series of courses, certificate options, graduate degree programs, and the establishment of research centers. Conversely, disciplines such as professional flight programs, aeronautical engineering technology, and aviation management still typically have traditional content-based plans of study that focus on the technical skills required for certification, and the curriculum has largely not been changed over the past few decades.

We recognize two primary options for incorporating AI into transportation higher education curricula. One possibility is to modify the existing curriculum or create new degree programs, including courses that may lead to students' understanding of the ethics, coding, data, and optimization of algorithms similar to what can be found in computer science programs. Adding new courses requires an understandable, rigorous bureaucratic process for most universities.

Conventional university programs without advanced AI course offerings that connect to transportation will have to consider if curriculum changes are needed to keep up with the rapid advancement of technology. Collegiate programs can outsource AI courses to other disciplines by amending the list of electives, thematic selectives, or other course categories. Another way to modernize curricula is by altering the required courses for a specific degree program or developing entirely new degree offerings. Either strategy will result in considerable effort and resources.

An alternative approach to incorporating AI into transportation higher education is the integration of end-user hardware and software into current courses. End-user products, which are becoming more available, can be obtained through equipment purchases and used in classrooms and labs. Ideally, modifying curriculum and implementing end-user technology can be utilized in parallel to enhance learning outcomes. Adoption of any of these strategies is not always a simple pathway forward.

The catalyst for change can come from various sources, including industry changes, technological advancements, administrative missions, the need for financial viability, etc. Involved parties may include trustees, presidents, provosts, deans, department chairs, committees, faculty senates, industry advisory boards, and catalog editors, among many other stakeholders. Research, data sharing, and communication throughout the organization are essential for a successful undertaking (Logue, 2018). Additionally, justifying the allocation and reallocation of resources will be necessary to achieve the buy-in from all stakeholders. Administrators and faculty should periodically examine the government's funding schemes and projects that promote improvements to education. Government organizations may provide grants, subsidies, or other rewards to organizations that actively strive to raise the caliber and applicability of their curricula. This can be beneficial for all university programs.

Moreover, universities must conduct small and/or large-scale market analyses to determine the need and develop standards, course outcomes, rubrics, syllabi, content, formative, and summative assessments. Further, mandated approval from government educational agencies and different levels of accreditation may be required and may take several years for full recognition.

Beyond the logistical challenges of modifying or developing a degree program, another challenge is the upskilling of faculty. Computer science fundamentals may not be under the purview of some faculty in various disciplines. It will take motivated and enthusiastic faculty with the time to engage in experiential learning, research, and additional formal or informal courses. Davis' Technology Acceptance Model (TAM) provides insight into this phenomenon. One's attitude toward adopting innovative technology or equipment is influenced by the perceived usefulness and ease of use of the product (Davis, 1989). Both factors are subjective in nature and affected by internal and external variables. If the adoption of new technology has a positive benefit, such as personal promotion or financial stability for an organization, those advantages will encourage the integration of modern technology (Davis, 1989).

Increasingly, assistance is available to instructors who are not proficient with this innovative technology. Instructors can leverage AI-infused end-user technologies, such as "off-the-shelf" simulators and sophisticated computer programs, to teach the fundamentals and provide examples of application and ethics. Technology companies, industry, and even institutions of higher education are increasingly offering AI training courses to educators that can be used to enhance pedagogy.

AI is not unlike other disruptive educational technologies that were previously introduced to the higher education landscape. Acceptance has been influenced by how well the technology fits into the existing pedagogical paradigm (Pence, 2019). For example, incorporating AI has created controversy among many instructors in the classroom due to, in part, its sudden presence across all disciplines. Following the release of Chat Generative Pre-Trained Transformer (ChatGPT) by OpenAI in November 2022, long-standing models of instruction and assessment were scrutinized. While this was not the first disruptive education technology introduced in the classroom, ChatGPT's access by students and ability to avoid detection by instructors cast doubt upon whether time-honored written assignments were the most pertinent technique for measuring learning. Some have questioned the relevance of such assignments if a non-sentient software application can create a work that is indiscernible from that written by an academically engaged student (Henriksen et al., 2023).

When AI is utilized as a tool for academic dishonesty, the concern must be addressed by the higher education community, as it would for any other instance of cheating. However, the tool should not be confused with the behavior. Just as higher education institutions have strived to create ethical graduates who can navigate their work environments, they must continue to prepare students who will be entering AI-infused careers (Pence, 2019).

Attention focused on the misuse of AI in higher education obscures the tremendous benefits and potential that it brings to academia. From a pedagogical perspective, when used correctly, ChatGPT or similar programs offer the possibility to personalize learning outside the classroom, with lessons tailored to individual learning styles. Moving content delivery outside the classroom creates space for instructors to use their dedicated time with students in other more creative and experiential learning opportunities (Rudolph et al., 2023). These activities can target higher order thinking skills, associated with the more advanced levels of Bloom's taxonomy.

While a product such as ChatGPT may be readily accessible, for aviation training in higher education, revolutionary AI products will have higher associated costs. Popular applications include enhanced flight simulators, with improved fidelity and immersive training options, which allow better translation of skills into the actual aircraft. Instruction delivered with the help of AI, as opposed to a singular flight instructor, provides the opportunity for more objective and thorough training. Eye-tracking technology can improve awareness of a pilot's thought process and decision-making tendencies. While the training community is still discovering the many applications of these products, their contributions to improved safety are already measurable. However, as discussed with the challenges of amending curricula, the costs associated with these newer technologies can present a strain on resources for many institutions of higher education.

2.1.1 Contemporary Examples of Artificial Intelligence Being Integrated into Higher Education

AI and other technical developments are driving a major revolution in higher education. The presented case studies showcase how the Southeastern Conference (SEC), Arizona State University (ASU), and Purdue University (three higher education institutions located in the United States) have resourced the implementation and inclusion of AI to enhance curriculum, research, and partnerships.

Southeastern Conference (SEC) AI Consortium Case Study

The purpose of the SEC AI Consortium is to expand prospects in the rapidly evolving domains of AI and data science, which are anticipated to serve as the bedrock for the future of industry, education, and research. Through this endeavor, SEC institutions are also responding to the requests made by local, state, and federal authorities who acknowledge the significance of improved training and workforce development (Southeastern Conference, 2021).

The SEC AI Consortium facilitates the sharing of educational resources among SEC universities, including curricular materials, certificate and degree program structures, and online presentations of seminars and courses. It also promotes workshops and academic conferences for faculty, staff, and students and aims to establish collaborative partnerships with industry. Furthermore, colleges will exchange optimal methods for ensuring students acquire AI and data science abilities that are advantageous to the local area and nation while also being relevant worldwide (Southeastern Conference, 2021).

ARIZONA STATE UNIVERSITY (ASU) CASE STUDY

ASU has emerged as a leader by using AI in its aviation program. The objective of this program is to improve Air Traffic Control (ATC) systems by utilizing AI technologies (Arizona State University, 2023). ASU's interdisciplinary programs integrate the fields of engineering, computer science, and data analytics to educate students in the practical use of AI in transportation systems. ASU collaborates with prominent companies to provide students with hands-on opportunities to work on projects and internships. This enables students to gain practical experience in implementing AI solutions for traffic control, autonomous vehicles, and urban mobility. ASU's endeavors are actively contributing to the progress of AI-powered ATC systems, which hold the potential to fundamentally transform the efficiency, safety, and sustainability of air transportation (Arizona State University, 2023).

PURDUE UNIVERSITY (PU) CASE STUDY

Purdue University has initiated a series of endeavors to incorporate AI into several facets of transportation. These projects involve conducting research, providing education, and collaborating with industry partners. Purdue's AI program objectives seek to tackle intricate difficulties, such as streamlining logistics networks, boosting public transit systems, and improving safety through predictive analytics. PU employs various interdisciplinary methods to provide students with the necessary abilities to address new challenges in AI for transportation (Purdue University, 2024).

In the area of autonomous vehicles, PU, in collaboration with Michigan University and the Department of Transportation, has created a Center for Connected and Automated Transportation (CCAT). The CCAT functions as a central point for multidisciplinary research and cooperation, bringing together faculty, students, and industry partners to promote the progress of AI-powered autonomous cars. Researchers in this transportation sector are examining autonomous vehicle systems, including driver perceptions, decision-making, and control (Purdue University, 2022). PU has begun enrolling students in a novel degree program and has created a research facility specifically focused on AI. The planned AI degree program is expected to provide students with extensive instruction in AI technologies, encompassing machine learning, deep learning, natural language processing, and robotics. The program aims to provide graduates with the ability to tackle intricate difficulties in different sectors by combining theoretical knowledge with practical applications (Purdue University, 2024).

Furthermore, alongside the degree program, investment in additional research facilities is providing a central location for interdisciplinary cooperation. This is uniting faculty, researchers, and industry partners in their effort to push the boundaries of AI technology. The primary objective is to foster innovation in AI applications, including fields such as healthcare, autonomous systems, cybersecurity, and smart cities, through advanced research initiatives and collaborations (Purdue University, 2024).

2.1.2 SUMMARY

The authors examined and assessed the challenges and potential advantages associated with higher education in transportation. Every transportation higher education program will inevitably encounter the difficulty of acquiring sufficient resources and funding to implement substantial improvements promptly. Nevertheless, higher education administrators must help enthusiastic advocates for curriculum change and hesitant faculty members accomplish the objectives of curriculum revision.

REFERENCES

Arizona State University. (2023). Revolutionizing Air Traffic Control Using AI. https://news.asu.edu/20231019-revolutionizing-air-traffic-control-using-ai

Davis, F. D. (1989). Perceived usefulness, perceived ease of use, and user acceptance of information technology. *MIS Quarterly*, 319–340.

Henriksen, D., Woo, L. J., & Mishra, P. (2023). Creative uses of ChatGPT for education: A conversation with Ethan Mollick. *TechTrends*, 1–6.

Logue, A. (2018). *Pathways to reform: Credits and conflict at the City University of New York*. Princeton University Press. https://chronicle.com/article/considering-curriculum-reform-better-do-your-homework-first/

Pence, H. E. (2019). Artificial intelligence in higher education: New wine in old wineskins? *Journal of Educational Technology Systems*, 48(1), 5–13.

Purdue University. (2022). Sustainable Transportation Systems https://engineering.purdue.edu/STSRG/research/CCAT/P_CCAT

Purdue University. (2024). Artificial Intelligence (College of Science). https://www.admissions.purdue.edu/majors/a-to-z/artificial-intelligence-science.php

Rudolph, J., Tan, S., & Tan, S. (2023). ChatGPT: Bullshit spewer or the end of traditional assessment in higher education? *Journal of Applied Learning & Teaching*, 6(1). https://doi.org/10.37074/jalt.2023.6.1.9

Southeastern Conference. (2021, November 30). SEC Universities Agree to Artificial Intelligence, Data Science Consortium. https://www.thesecu.com/news/sec-universities-agree-to-artificial-intelligence-data-science-consortium/

2.2 THE ROLE OF AI IN AI TRAINING/OPERATIONS

Dimitrios Ziakkas, Konstantinos Pechlivanis, and Anastasia Kosmidou

"Artificial Intelligence Training and Operations" delves into the transformative impact of AI across various industries, focusing on its revolutionary applications in training, operations, weather forecasting, route optimization, maintenance, and human factors management. AI's integration into training and operations is highlighted as an enhancement and a transformative force, offering adaptive learning environments, and optimizing decision-making processes to improve operational efficiency and reduce human error (Ziakkas et al., 2023). The seamless incorporation of AI into transportation curricula represents a forward-thinking approach to education that aligns with industry needs and technological advancements. Additionally, AI's ability to manage human factors such as fatigue risk, human error, and workload distribution is emphasized. AI is posited as a catalyst for innovation and optimization in training and operational practices, setting new standards for efficiency, safety, and effectiveness.

2.2.1 MENTAL REHEARSAL IN AVIATION AND EMERGING TECHNOLOGIES: THE STEP FORWARD IN TRAINING

Every aviator comes across the term "chair flying" at the onset of his/her career – a concept hard to grasp but an invaluable and irreplaceable tool for life. Along the years, the development of mental models was augmented with devices of continuously increasing complexity. Since the pioneering Link Trainer until the rise of Full Flight Simulators, simulators have revolutionized pilot training by providing accurate reproductions of specific aircraft cockpits with full-motion features over the years. With the advent of Virtual Reality (VR) and Augmented Reality (AR) in recent decades, aviation training has entered a new dimension, offering immersive and cost-effective experiences. The present era sees the emergence of Extended Reality (XR) simulators, combining the strengths of VR and AR to provide realistic simulations, enhance engagement, and offer immediate feedback, thereby continuing the rich legacy of advancing pilot training and aviation safety.

Mental rehearsal stands as a cornerstone in the realm of pilot training and the sustained honing of aviation skills. Piloting aside, its significance lies in an array of benefits that elevate individual performance to new heights with proven records in ATC, sports, and healthcare. Firstly, mental rehearsal becomes a simulated haven for decision-making prowess. Pilots immerse themselves in a risk-free environment, adeptly navigating diverse flight scenarios and mastering emergency procedures, fostering a heightened ability to respond promptly and effectively in the face of unforeseen challenges. Secondly, the cognitive terrain expands to include enhanced situational awareness. Through mental rehearsal, pilots traverse the intricacies of airspace navigation, system monitoring, and communication with ATC, cultivating an acute awareness crucial for safe and precise flight operations. Beyond the technical facets, mental rehearsal emerges as a psychological ally for pilots, offering sanctuary against stress and anxiety. By providing a platform for the practice of coping strategies, it becomes a pre-emptive measure to alleviate the cognitive load during actual flight operations. The mental stage becomes a rehearsal space not only for decisions and maneuvers but also for emotional resilience.

2.2.2 Benefits of Extended Reality (XR) in Pilots' Mental Rehearsal

Emerging technologies, such as XR, can significantly augment mental rehearsal in aviation by offering advanced tools and platforms for more immersive and effective training (Dymora et al., 2021; Flores et al., 2023). Here are some key ways in which these technologies enhance mental rehearsal for pilots:

Realistic Simulation: XR technologies, particularly VR, provide highly realistic and immersive flight simulations.

Enhanced Engagement: XR technologies capture the attention of pilots and enhance their motivation to engage in mental rehearsal. Additionally, XR provides an engaging, interactive, and captivating learning environment that encourages active participation.

Immediate Feedback: XR applications can provide immediate feedback during mental rehearsal sessions. This real-time feedback mechanism also accelerates the learning process, as pilots can make corrections and reinforce correct behaviors more effectively.

Cost-Effective Training: Traditional flight training can be costly, involving expenses such as fuel, aircraft maintenance, and physical simulators. XR-based mental rehearsal significantly reduces these costs. While initial investments in XR technology may be required, the long-term benefits in terms of cost savings are substantial. Virtual environments can be adapted for various training scenarios, reducing the need for physical equipment and the expenses associated with traditional training. Additionally, XR training can be easily updated and customized to meet the specific needs of different pilots and aircraft types.

Flexible and Remote Training: XR-based mental rehearsal is adaptable to different training scenarios and can be conducted remotely. Pilots can engage in mental rehearsal from their own locations and at their convenience, eliminating the need for on-site training facilities.

To conclude, mental rehearsal is an indispensable aspect of aviation training and pilot performance. It enables pilots to make better decisions, enhance situational awareness, reduce stress, develop muscle memory, and maintain their skills over time. Emerging technologies like XR have brought a new dimension to mental rehearsal, providing realistic simulation, enhanced engagement, immediate feedback, cost-effective training, and flexibility.

2.2.3 ARTIFICIAL INTELLIGENCE IN SHIPPING OPERATIONS
Michail Diakomihalis

Many companies in the shipping industry have put in place systems that evaluate AI performance and have established specific indicators such as Key Performance Indicators, through which the performance of the business in various areas and functions is monitored, and their coordination with the company's strategy. Relevant indicators have been created to monitor and evaluate the functions of financial management, operation, safety, human resources, maintenance, repairs, etc., to inform the management and officials of the vessel about the actual situation of the ship and the company. The operation of the existing system can include data from employee analytics to help evaluate, control, and emerge new ideas related to the optimal use and evaluation of employees' time. In this way, employees will be able to control their performance and have access to the personal data generated by the system so that they can improve their performance, while managers will be able to make more effective decisions for the achievement of the company's goals (Lester, 2019).

Human resource data analytics should be used from the start, with care taken to protect the privacy of some data, while fully understanding how powering applications within the context of artificial intelligence (AI) might lead to changes in company operations. According to the International Safety Management (ISM) Code, the shipping company must appoint a responsible person from among its executives to work in the office of the company's general management, with the sole responsibility of monitoring and ensuring the safe operation of its ships, as well as serving as the connecting link of communication between the company and its ships. As the Designated in the Safety Management System, this executive is granted direct access to the upper echelons of management within the maritime company. Furthermore, the shipping company must precisely establish the inter-worker relationships that must be maintained. The Designated Person in Safety is responsible for implementing and complying with the principles and objectives of Security Management.

The main pillars of the IMS Code are: (i) the protection of human life through preventing human accidents and loss of life, (ii) the safeguarding of the marine environment; however, extending to the whole environment, (iii) maintaining and optimizing the technical and operational efficacy of ships, (iv) the protection of the ship and the crew from any illegal action, and (v) the management of ships is always carried out safely and efficiently, as well as ensuring the safe transport of passengers.

The ISM Code requires the Safety and Training Department to create the Safety Management System. It assists the General Management of the Company in Safety and Environmental Protection matters by reporting malfunctions and deficiencies and suggesting ways to improve procedures through internal audits and recording practices that violate company policy. The shipping department general manager monitors and encourages SMS personnel to participate in the implementation of Safety and Environmental Protection regulations and audits the Safety Management System regularly. The Safety and Training Manager can sign and bind the firm for any department-related product supply transaction up to the General Manager's approval. To summarize, the Director of the Safety & Training Department is responsible for:

- Supervise the implementation of the regulations in the SMS manual through the Technical Department. Keeping records and reports on security issues.
- Supervise compliance with Safety and Environmental Protection rules and check the completion of the relevant forms sent to the Company's Directorate.
- Investigate and analyze accidents and incidents related to Safety and Environmental Protection and propose alternative actions to prevent them.
- Process the topics of the conferences on the ships and the programs of exercises and readiness.
- Control navigational operations and safety-related maintenance.
- Check ships' logbooks and ensure regular updates from publications such as "British Admiralty Notices and To Mariners."
- Define the conduct of safety and SOPEP exercises, review and analyze their results, and estimate the stocks of safety equipment monthly.
- Inspect ships at least quarterly for compliance with safety and health rules, and propose and issue newsletters regarding the above and ship obligations.
- Monitors and takes care of the validity of the certificates for the ship's category (Hull-Safety-Load Line).
- Preparation for repairs and tanking, suggesting ways, and subcontracting for the repairs and to take care of deck supplies for the safe operation of the ship.

Supervised learning, a well-known and accepted modern form successfully applied in shipping companies, has proven to facilitate the management of companies, ships, and employees. It is most prevalent in the crew department, which is a shipping company's human resources department. The AI virtual assistant could save time by automatically scanning e-mail messages, copying their attachments, and filing them where they belong, having them available for future use, forwarding them to people in the business or other ships that may need them without the employment of a workforce, and with greater assurance of documentation and a lack of error.

REFERENCES

Dymora, P., Kowal, B., Mazurek, M., & Romana, S. (2021). The Effects of Virtual Reality Technology Application in the Aircraft Pilot Training Process. In *IOP Conference Series: Materials Science and Engineering* (Vol. 1024, p. 012099). IOP Publishing. doi: 10.1088/1757-899X/1024/1/012099.

Flores, A., Ziakkas, D., & Dillman, B. (2023). Artificial Cognitive Systems and Aviation Training. In T. Ahram, W. Karwowski, P. Di Bucchianico, R. Taiar, L. Casarotto & P. Costa (eds). *Intelligent Human Systems Integration (IHSI 2023): Integrating People and Intelligent Systems. AHFE (2023) International Conference. AHFE Open Access* (Vol. 69). AHFE International, USA. https://doi.org/10.54941/ahfe1002838.

Lester, J. F. (2019). Applying Artificial Intelligence in the Maritime Shipping Industry: Sorting through the Digital Deluge, Facilitating Decision Making, and Freeing Personnel to Focus on More Important Tasks. https://scholarworks.calstate.edu/downloads/r494vm08r

Ziakkas, D., Vink, L.-S., Pechlivanis, K., & Flores, A. (2023). *Implementation Guide for Artificial Intelligence in Aviation: A Human-Centric Guide for Practitioners and Organizations.* ISBN 9798863704784.https://www.amazon.com/IMPLEMENTATION-GUIDE-ARTIFICIAL-INTELLIGENCE-AVIATION/dp/B0CL7Y4HPJ

2.3 THE ROLE OF AI IN WEATHER PREDICTION, PLANNING, ROUTE OPTIMIZATION, AND SCHEDULING

Anastasios Plioutsias and Dimitrios Ziakkas

2.3.1 AI IN WEATHER PREDICTION

The advancement in weather forecasting through the integration of AI marks a significant progression in our comprehension and prediction of meteorological phenomena. Traditionally, weather prediction was predominantly based on statistical models, the principles of physics, and direct observations. These conventional approaches, although fundamental, frequently encountered difficulties in coping with the intricate and unpredictable nature of atmospheric systems, thereby constraining their forecast precision, particularly concerning extreme meteorological events.

The introduction of machine learning (ML) and deep learning techniques has significantly transformed weather forecasting. Unlike traditional methods, AI technologies can gather and analyze vast amounts of data from various sources, including satellites, radars, terrestrial weather stations, and marine buoys. This allows them to identify patterns and accurately predict future weather conditions. These advanced models refine their predictive capabilities by learning from historical weather data, continually improving their accuracy as they integrate new information. One of the main advantages of AI-based models is their ability to execute high-resolution simulations. TIBM's Deep Thunder project is an excellent example of how AI can combine with hyper-localized weather models to provide accurate weather forecasts. This initiative shows how AI can help improve operational strategies and disaster response measures. Google and the National Oceanic and Atmospheric Administration (NOAA) have joined forces to improve weather predictions using AI and ML. NOAA will use Google's advanced computing resources and AI expertise to enhance its predictive models for severe weather events. The European Centre for Medium-Range Weather Forecasts (ECMWF) is using AI to improve the accuracy of its worldwide weather forecasts. ECMWF has significantly improved its predictive accuracy by incorporating AI into data assimilation techniques, especially for extreme weather conditions.

2.3.2 AI IN PLANNING AND DECISION MAKING

Integrating AI into planning and decision-making frameworks is crucial in navigating the complex repercussions of weather across diverse sectors. Incorporating accurate weather forecasts into strategic planning empowers businesses and governmental entities to make well-informed choices, mitigate potential risks, and fine-tune their operations. AI technologies are instrumental in this endeavor, as they efficiently process and analyze extensive weather data, enabling precise condition forecasts. These predictive insights are crucial for a wide array of planning initiatives, ranging from scheduling in agriculture and construction to logistics in supply chain management and event coordination, thereby significantly reducing weather-associated risks (Boukabara, 2022).

Similarly, AI can forecast likely weather-induced delays in the construction industry, prompting pre-emptive scheduling adjustments to circumvent financial excesses and project postponements. AI-driven systems use weather forecasts to advise farmers on the best times for planting, irrigating, and applying pesticides, thus maximizing crop yields and reducing waste. Construction project managers use AI to adjust schedules based on weather predictions, ensuring worker safety and project efficiency. AI optimizes shipping routes and schedules for logistics companies to account for weather conditions, reducing delays and protecting goods. These applications highlight AI's capacity to dynamically adapt operations in response to weather predictions, leading to more efficient and cost-effective outcomes.

2.3.3 ROUTE OPTIMIZATION

Integrating AI into route optimization significantly advances logistics and transportation efficiency, mainly when dealing with variable weather conditions. Dynamic route optimization is crucial for industries that rely on timely deliveries and efficient transportation routes, such as logistics, shipping, and passenger transport. Weather events can cause severe disruptions to these routes, leading to delays, increased operational costs, and compromised safety. AI algorithms integrate and analyze vast datasets, including real-time and forecasted weather information, traffic conditions, and road infrastructure data, to determine the most efficient routes (Boukabara, 2022). These algorithms use predictive models to assess the impact of various weather conditions on transportation, such as reduced visibility, slippery roads, or potential flooding, and then calculate the best possible routes to avoid or minimize these impacts.

The process involves several steps:

Data Collection: Gathering real-time weather data and forecasts from meteorological sources.

Data Analysis: Analyzing the collected data to assess potential weather-related disruptions.

Route Simulation: Simulating various routing scenarios to identify the most efficient paths considering the predicted weather conditions.

Optimization: Employing optimization algorithms to select the route that minimizes travel time, costs, and safety risks.

AI-optimized routes can reduce travel time and fuel consumption by avoiding weather-affected areas and traffic congestion, leading to more efficient operations and lower carbon footprints. Minimizing delays and improving fuel efficiency directly reduce operational costs. Finally, the ability to anticipate and avoid weather-related disruptions prevents costly last-minute changes and penalties.

2.3.4 Scheduling Adjustments

AI use in scheduling adjustments is crucial for improving operational resilience and efficiency, particularly in industries where timing and precision are paramount. AI systems are highly proficient in analyzing vast amounts of data, including historical weather patterns, real-time weather updates, and operational performance metrics to predict potential disruptions. Such models use predictive analytics and ML algorithms to accurately anticipate delays, cancellations, and other disruptions, allowing for proactive scheduling adjustments. By predicting these disruptions, organizations can minimize their impact, ensuring smoother operations and increased reliability. After identifying possible disruptions, AI systems can suggest the best possible scheduling adjustments (Ziakkas et al., 2023).

Airlines and shipping companies use AI to dynamically adjust their schedules in response to weather predictions and other disruptions. AI helps these companies adjust their shipping routes and schedules based on weather forecasts and port availability, reducing the risk of delays and damage to cargo. AI-driven scheduling adjustments also ensure service continuity and safety during adverse weather conditions by adjusting routes and schedules to maintain optimal service levels and passenger safety.

2.3.5 AI in Weather Prediction, Planning, Route Optimization, and Scheduling: The Shipping Case Study

NATALIA MARCOVICI

AI has transformed numerous industries, including the transportation business. AI has become a transformative force for ships and the maritime sector, enhancing efficiency, safety, and enabling autonomous operations. AI has significantly improved productivity, safety, and sustainability in the maritime sector by processing large data sets, making instant judgments, and optimizing operations. The recent case study examines the various uses of AI in shipping, including autonomous ships, predictive maintenance, and route optimization.

AI is widely used in shipping to create autonomous or unmanned vessels. The warships utilize sophisticated sensors, ML algorithms, and computer vision to autonomously traverse the oceans. This not only decreases the likelihood of human mistakes but also creates opportunities for saving costs and improving operational efficiency. Autonomous ships can function continuously, without crew breaks, which could accelerate delivery schedules and save transportation expenses. AI-powered predictive maintenance has significantly changed the shipping sector. AI systems can forecast potential failures of vital ship components by examining previous data and current sensor data. This enables pre-emptive maintenance, decreasing the likelihood of unforeseen malfunctions, saving downtime, and prolonging the lifespan of equipment. By continuously adapting to changing circumstances, these systems optimize fuel efficiency, reduce environmental impact, and enhance overall operational performance. Furthermore, AI plays a pivotal role in improving the accuracy and efficiency of cargo handling at ports. Automated cranes, guided by AI algorithms, can precisely load and unload containers from ships, speeding up the process and reducing the likelihood of errors. AI-powered image recognition systems also enhance security by identifying and tracking cargo, ensuring compliance with customs regulations, and preventing illicit activities.

In the realm of safety, AI contributes significantly to risk mitigation. Advanced sensor technologies, such as radar and lidar, coupled with AI algorithms, enable vessels to detect and respond to potential hazards in real-time. These systems can autonomously navigate around obstacles, avoid collisions, and ensure the safety of both the ship and its surroundings. In addition, AI-driven predictive analytics can assess the likelihood of accidents based on historical data, allowing for the implementation of preventive measures. The shipping industry is notorious for its paperwork-intensive processes related to customs, documentation, and compliance. AI streamlines these operations by automating document verification, ensuring accuracy, and reducing the time and resources required for bureaucratic tasks. Moreover, AI facilitates the monitoring and enforcement of environmental regulations in the shipping sector. By analyzing data on fuel consumption, engine efficiency, and other relevant parameters, AI contributes to the development of sustainable practices, aligning the industry with global environmental goals. In conclusion, AI has ushered in a new era for the shipping industry, transforming traditional practices and unlocking unprecedented possibilities. As technology continues to evolve, the shipping sector can expect further advancements that will not only improve operational efficiency and cost-effectiveness but also contribute to a more sustainable and environmentally conscious future.

REFERENCES

Boukabara, S.-A. (2022, March 1). Outlook for exploiting artificial intelligence in the earth and environmental sciences. *Bulletin of the American Meteorological Society*. https://www.academia.edu/en/72704113/Outlook_for_Exploiting_Artificial_Intelligence_in_the_Earth_and_Environmental_Sciences

Ziakkas, D., Vink, L.-S., Pechlivanis, K., & Flores, A. (2023). *Implementation Guide for Artificial Intelligence in Aviation: A Human-Centric Guide for Practitioners and Organizations*. ISBN 9798863704784.

2.4 AI-POWERED MAINTENANCE AND PREDICTIVE ANALYSIS FOR VEHICLE HEALTH MONITORING AND MAINTENANCE SCHEDULING

Ioannis Katsidimas and Athanasios Kotzakolios

AI is significantly transforming aviation maintenance, presenting innovative approaches to enhance reliability, safety, and operational efficiency. Recent studies illustrate AI's application across various facets of maintenance, such as employing Artificial Neural Networks for Engine Health Monitoring, facilitating predictive maintenance through AI-supported strategies, and integrating Computer-Based Training systems for improved maintenance training (Pınar, 2022). Furthermore,

Artificial Intelligence Training and Operations

AI's incorporation in the design of functionalized composite airframe structures and the use of AR for maintenance training exemplify the shift towards intelligent, self-diagnosing systems and advanced training methodologies (Ye et al., 2005). The digital transformation in maintenance operations, driven by Industry 4.0 technologies, underscores the pivotal role of AI in advancing airworthiness, leading to safer, more reliable, and cost-effective maintenance processes. Collectively, the advances in Big Data analytics, Predictive Maintenance Frameworks, Big Data analytics, Predictive Maintenance Frameworks, ML, and Digital Twins together unlock AI's potential to revolutionize aviation maintenance, improving aircraft diagnostics, efficiency, and flight safety.

2.4.1 BIG DATA ANALYTICS IN AVIATION MAINTENANCE

The aviation industry is increasingly leveraging Big Data Analysis for predictive maintenance, revolutionizing the way maintenance operations are conducted and enhancing overall operational efficiency and reliability. Predictive maintenance, which relies on continuous data collection, monitoring, and the application of advanced analytics, promises a future where aircraft parts can preemptively signal their need for repair or replacement, thus ensuring higher safety standards and operational efficiency.

Combining cloud-based analytics with industrial machinery has led to the emergence of the "Industrial Internet," creating new possibilities for improving productivity. This technological convergence enables the gathering and examination of large amounts of data from electronic sensors integrated into industrial machines, such as aircraft engines and systems. Advanced analytics provide insights that optimize the functioning of machines, engine fleets, and complex systems like airplanes and airports in more effective ways. The comprehensive data analysis not only predicts the maintenance needs but also optimizes supply chain efficiencies, contributing to a more reliable and efficient aviation sector.

Further advancements in ML models, specifically Long Short-Term Memory (LSTM) neural networks, have proven effective in predicting the replacement times for aircraft system components, thereby optimizing benefits and reducing maintenance costs. These models process and predict significant events over long intervals and delays in time series, offering a more accurate estimation of maintenance needs, which in turn, ensures flight safety and operational efficiency. Moreover, hybrid ML approaches that combine natural language processing techniques with ensemble learning have been developed to predict extremely rare aircraft component failures. Such models have proven to outperform existing methods in precision, recall, and f1-score, demonstrating the potential of ML in transforming predictive maintenance practices (Dangut et al., 2020).

Deep learning techniques, particularly the application of Long Short-Term Memory network-based Auto-Encoders (LSTM-AE), have been employed for fault detection and classification in complex aircraft systems. These methods use raw time-series data from different sensors to create a reconstruction model that is based on the healthy system behavior. The system's health indicators and fault feature vectors obtained from the reconstruction error matrix enable effective fault troubleshooting, highlighting deep learning's role in enhancing predictive maintenance capabilities (Ning et al., 2021).

The successful implementation of predictive data analytics in aviation maintenance is not solely dependent on technological advancements but also requires a cultural shift within organizations. The aviation maintenance community must embrace data-driven analysis and predictive solutions to fully benefit from these advancements. This involves not only understanding the technological aspects but also addressing cultural barriers and fostering an environment that supports innovation and trust in predictive maintenance systems (Wilson et al., 2022).

2.4.2 Machine Learning for Failure Prognostics

The application of ML in the domain of failure prognostics for aircraft is witnessing substantial advancements, particularly through the deployment of sophisticated models like ProgNet. This model exemplifies the potential of adaptive deep learning by offering a significant boost in the prognosis capabilities for aircraft engine damage under the fluctuating conditions of real flights. Its effectiveness is underscored by remarkable performance metrics on the comprehensive N-CMAPSS dataset, which simulates a variety of flight scenarios, thereby demonstrating its robustness and adaptability to real-world applications (Berghout et al., 2022).

In parallel, Convolutional Neural Networks (CNNs) are being leveraged to explore engine gas path prognostics, with studies revealing their superior feasibility and effectiveness in diagnosing and predicting engine conditions when compared to traditional data-driven methods. The use of CNNs allows for the extraction of complex patterns from sensor data, providing a more nuanced understanding of engine health (Jiang et al., 2019).

Furthermore, the integration of bi-directional Long Short-Term Memory (BLSTM) networks with transfer learning methodologies is enhancing the accuracy of Remaining Useful Life (RUL) predictions. In cases where failure progression data is scarce, this approach allows the use of knowledge from related datasets to improve prediction models, overcoming one of the biggest challenges in data-driven prognostics (Zhang et al., 2018).

The synergy between physics-of-failure modeling and diagnostic information is another area of noteworthy progress. By combining detailed failure mode analysis with real-time sensor data, models can now offer more accurate prognostics, minimizing the inherent uncertainties associated with complex machinery operations, as demonstrated in studies involving H-60 helicopter gear cases (Kacprzynski et al., 2004).

Innovations such as adaptive denoising techniques and degradation pattern learning are further refining the precision of RUL predictions. Adaptive denoising helps in enhancing the signal-to-noise ratio of sensor data, thereby improving the input quality for prognostic models. Degradation pattern learning, on the other hand, employs neural networks to identify and learn the progression patterns of wear and tear in engine components, making the prediction of RUL more accurate (Zhao et al., 2017).

Moreover, hybrid data preparation models are being developed to enhance the accuracy of failure predictions. By combining feature selection techniques and data elimination methods, these models can effectively filter out irrelevant or noisy data, thereby refining the inputs for ML algorithms. This approach has shown success in predicting equipment failures, highlighting the potential for more reliable maintenance planning (Celikmih et al., 2020).

Ensemble learning-based approaches are also being explored for their potential to combine predictions from multiple models, thereby increasing the robustness and accuracy of RUL estimates. By leveraging the strengths of various predictive models, such as random forests, neural networks, and regression models, ensemble methods can offer a more nuanced and accurate prediction of aircraft engine RUL, outperforming many standalone predictive models (Li et al., 2018).

Automated Visual Inspection (AVI) from the MLEAP interim public report discusses the use of AVI in the aviation industry, specifically in aircraft maintenance. AVI uses deep learning and computer vision to visually inspect aircraft surfaces for defects, such as lightning strike impacts and dents. This technology aims to assist human inspectors by reducing the inspection duration and improving reliability, as human inspections can be subjective and error prone.

The report highlights the development of a system by Airbus for in-service damage detection, focusing on diagnostic assistance to reduce maintenance duration, automatic detection and classification of damages, and leveraging a combination of images and videos for this purpose. The approach involves using a Siamese network architecture, built upon pre-trained VGG16 networks, for multitasking in damage detection and characterization. The implementation steps include using the OpenCV library for image stain detection, binary classification for stain detection, multi-class damage classification, and employing a cross-entropy layer and Adam optimizer for training, with a targeted performance of over 95% accuracy and recall.

However, the report also outlines several challenges associated with AVI, such as the high dimensionality of the use case, the need for complete and relevant data, the importance of diverse input data for model robustness, and the ongoing comparison of various models to address these challenges. Technical challenges include ensuring AVI systems' accuracy, robustness to detect a wide range of surface defects, scalability across different aircraft types, and effectiveness under various conditions.

2.4.3 Integrated Frameworks and Data-Driven Techniques for Predictive Maintenance

The integration of comprehensive predictive maintenance frameworks and advanced data-driven methodologies has significantly enhanced the efficiency, safety, and cost-effectiveness of aircraft maintenance systems. A groundbreaking study by Yan et al. (2020) proposed an integrated predictive maintenance framework, encompassing historical data analysis, system health assessment, remaining useful life prediction, and strategic maintenance decision-making. This framework has demonstrated the potential for substantial reductions in maintenance costs and improvements in mission reliability, marking a significant shift from traditional preventive maintenance approaches. Complementing this, the work of Baptista et al. (2018) showcased the integration of data-driven techniques with ARMA modeling to forecast fault events accurately and optimize maintenance schedules, thus enhancing predictive accuracy and operational efficiency.

Daily and Peterson (2017) emphasized the role of big data analysis in predictive maintenance, highlighting the promise of increased reliability and operational efficiencies through the integration of cloud-based analytics with industrial machinery. The predictive analytics framework, driven by the digital twin concept as outlined by Heim et al. (2020), further supports the detailed description and justification of maintenance routines, offering methods to describe the remaining useful life of aircraft parts through data sets and digital twin models. Moreover, Dangut et al. (2020) proposed a hybrid ML approach blending natural language processing techniques with ensemble learning, aiming to predict extremely rare aircraft component failures. This approach, tested on real aircraft maintenance data, demonstrated superior performance over traditional methods, addressing the challenge of data imbalance in predictive modeling. Khan et al. (2021) also contributed to this field by surveying recent trends and challenges in predictive maintenance of aircraft engines and hydraulic systems, emphasizing the importance of real-time data for diagnosis and prognosis.

Structural health monitoring (SHM) plays a vital role in decision-making concerning predictive maintenance operations and strategies, making it a very wide application subject with many challenges to apply sensing and processing capabilities. The core element of a modern digitalized SHM system is sensing, which in most cases includes vibroacoustic, piezoelectric, stain gauges, fiber optics, and other sensing elements. The most common signal processing methods include Fourier transformation (and its variants such as fast and short-time), statistical analysis over time series, Cohen's class, wavelet transform, Hilbert–Huang transformation, neural networks (NN), Bayesian classifiers, and further hybrid approaches. Finally, there are typically two approaches that mainly attempt to deal with SHM problems: (i) knowledge-based or model-driven models, which in some cases either neglect realistic and real-life conditions or make some convenient assumptions, and (ii) data-based models, which suffer from sufficient data availability, as running an experiment multiple times is costly in terms of time, equipment and personnel. Integrated TinyML seems like a quite promising technology to address some of these challenges in a transparent and decentralized way, offering technical assistance in decision-making using processed information that is produced autonomously and without further conflicts and impedances. TinyML can process this data directly on the sensor or nearby processors, reducing the need for data transmission to larger systems and enabling quicker response times. Additionally, since TinyML operates on low-power devices, it can perform essential computations without significantly draining power resources. Katsidimas et al. (2023) present a novel system designed for structural health monitoring (SHM) using TinyML on low-cost, resource-constrained IoT devices. The study focuses on detecting and localizing impacts on a thin plate made of polymethyl methacrylate (PMMA) using piezoelectric transducers and ML models like Random Forest and Shallow Neural Network, achieving over 90% accuracy in real-time impact localization with less than 400 ms latency. The research introduces a novel dataset for impact events and discusses the methodology, data collection, processing, and ML models used in the experiment. The use of PMMA as an impact material is important, since it is widely used in aircraft windows and canopies, exterior lighting and lenses, or even small aircraft windshields. The dataset and the experimental design are publicly available in the Zenodo database (https://zenodo.org/records/7199346).

2.4.4 Digital Twins

The concept of Digital Twins, a term first envisioned by NASA, represents a significant leap in how we manage and maintain complex systems such as aircraft. NASA's vision for Digital Twins was to create a synchronized digital replica of a physical object or system, spanning its lifecycle, and using real-time data to enable prediction, optimization, and simulation capabilities. This paradigm shift, aimed at addressing the extreme requirements of future NASA and U.S. Air Force vehicles, involves integrating ultra-high-fidelity simulation with the vehicle's onboard health management system and historical data to ensure unprecedented levels of safety and reliability. In the realm of aircraft maintenance, Digital Twins can deliver a transformative approach by providing a dynamic and up-to-date representation of the aircraft throughout its operational life. This not only includes the visualization of the aircraft in its current state but also the integration of predictive analytics to foresee potential maintenance issues before they occur, thereby reducing downtime and increasing operational efficiency.

A core component of Digital Twins is the integration of physics-based models, which are crucial for accurately simulating and predicting the behavior of complex systems like aircraft. These models, grounded in the fundamental principles of physics, ensure that the digital twin can reliably mirror the physical twin under various conditions and scenarios (Heim et al., 2020). The fusion of physics-based models with data-driven analytics, such as ML algorithms, further enhances the predictive capability of Digital Twins. This hybrid modeling approach enables the Digital Twin to adapt and update itself based on real-time data, ensuring the accuracy and relevance of the insights provided for maintenance and operational decision-making.

The integration of Digital Twins into aircraft maintenance and repair has been showcased through various innovative applications. For example, the use of AR combined with Digital Twins allows for the overlaying of a 3D model from the aircraft's Digital Twin directly onto physical components, enhancing inspection processes and enabling remote expert collaboration. This approach not only streamlines manual inspections but also facilitates efficient damage assessment and repair planning. Additionally, the development of a Digital Twin that models an aircraft's operational life cycle significantly improves the precision of MRO operations. By leveraging ML, this model enhances the prediction of defects and optimizes spare part planning, leading to more efficient and cost-effective maintenance strategies. Furthermore, the concept of an Airframe Digital Twin offers a tailored computational model for each aircraft, serving as a virtual health sensor to forecast future maintenance needs and assure structural integrity (MLEAP Consortium, 2023).

REFERENCES

Baptista, M., Sankararaman, S., Medeiros, I., Nascimento, C., Prendinger, H., & Henriques, E. (2018). Forecasting fault events for predictive maintenance using data-driven techniques and ARMA modeling. *Computers & Industrial Engineering*, 115, 41–53. https://doi.org/10.1016/j.cie.2017.10.033.

Berghout, T., Mouss, M., Mouss, L., & Benbouzid, M. (2022). ProgNet: A transferable deep network for aircraft engine damage propagation prognosis under real flight conditions. *Aerospace*. https://doi.org/10.3390/aerospace10010010.

Celikmih, K., Inan, O., & Uguz, H. (2020). Failure prediction of aircraft equipment using machine learning with a hybrid data preparation method. *Scientific Programming*, 2020, 8616039:1-8616039:10. https://doi.org/10.1155/2020/8616039.

Dangut, M., Skaf, Z., & Jennions, I. (2020). An integrated machine learning model for aircraft components rare failure prognostics with log-based dataset. *ISA Transactions*. https://doi.org/10.1016/j.isatra.2020.05.001.

Heim, S., Clemens, J., Steck, J., Basic, C., Timmons, D., & Zwiener, K. (2020). Predictive Maintenance on Aircraft and Applications with Digital Twin. In *2020 IEEE International Conference on Big Data (Big Data)* (pp. 4122–4127). https://doi.org/10.1109/BigData50022.2020.9378433.

Jiang, Z., Fang, H., Shi, H., Ren, S., Yang, H., & Wang, F. (2019). The prognostic method of engine gas path based-on convolutional neural network. *DEStech Transactions on Computer Science and Engineering*. https://doi.org/10.12783/DTCSE/ICITI2018/29126.

Kacprzynski, G., Sarlashkar, A., Roemer, M., Hess, A., & Hardman, B. (2004). Predicting remaining life by fusing the physics of failure modeling with diagnostics. *JOM*, 56, 29–35. https://doi.org/10.1007/S11837-004-0029-2.

Katsidimas, I., Kostopoulos, V., Kotzakolios, T., Nikoletseas, S. E., Panagiotou, S. H., & Tsakonas, C. (2023). An impact localization solution using embedded intelligence-methodology and experimental verification via a resource-constrained IoT device. *Sensors*, 23, 896. https://doi.org/10.3390/s23020896

Khan, K., Sohaib, M., Rashid, A., Ali, S., Akbar, H., Basit, A., & Ahmad, T. (2021). Recent trends and challenges in predictive maintenance of aircraft's engine and hydraulic system. *Journal of the Brazilian Society of Mechanical Sciences and Engineering*, 43. https://doi.org/10.1007/s40430-021-03121-2.

Li, Z., Goebel, K., & Wu, D. (2018). Degradation modeling and remaining useful life prediction of aircraft engines using ensemble learning. *Journal of Engineering for Gas Turbines and Power.* https://doi.org/10.1115/1.4041674.

MLEAP Consortium. (2023). EASA Research – Machine Learning Application Approval (MLEAP) Interim Technical Report. European Union Aviation Safety Agency. Horizon Europe Research and Innovation Programme Report, May 2023. https://www.easa.europa.eu/en/research-projects/machine-learning-application-approval

Ning, S., Sun, J., Liu, C., & Yi, Y. (2021). Applications of deep learning in big data analytics for aircraft complex system anomaly detection. *Proceedings of the Institution of Mechanical Engineers, Part O: Journal of Risk and Reliability*, 235, 923–940. https://doi.org/10.1177/1748006X211001979.

Pinar, Adem. (2022). Artificial intelligence supported aircraft maintenance strategy selection with q-Rung orthopair fuzzy TOPSIS method. *Journal of Aviation*, 6. https://doi.org/10.30518/jav.1150219.

Mrusek, B., Wilson, J., Solti, J., Reimann, M., Witcher, K. (2022). Predictive Data Analytics in Aviation Maintenance: A Cultural Perspective. In: Tareq Ahram, Waldemar Karwowski, Pepetto Di Bucchianico, Redha Taiar, Luca Casarotto and Pietro Costa (eds) Intelligent Human Systems Integration (IHSI 2022): Integrating People and Intelligent Systems. AHFE (2022) International Conference. AHFE Open Access, vol 22. AHFE International, USA. http://doi.org/10.54941/ahfe100988

Ye, L., Lu, Y., Su, Z., & Meng, G. (2005). Functionalized composite structures for new generation airframes: a review. *Composites Science and Technology*, 65, 1436–1446. https://doi.org/10.1016/J.COMPSCITECH.2004.12.015.

Zhang, A., Wang, H., Li, S., Cui, Y., Liu, Z., Yang, G., & Hu, J. (2018). Transfer learning with deep recurrent neural networks for remaining useful life estimation. *Applied Sciences*. https://doi.org/10.3390/APP8122416.

Zhao, Z., Liang, B., Wang, X., & Lu, W. (2017). Remaining useful life prediction of aircraft engine based on degradation pattern learning. *Reliability Engineering & System Safety*, 164, 74–83. https://doi.org/10.1016/j.ress.2017.02.007.

2.5 ARTIFICIAL INTELLIGENCE APPLICATIONS AS COGNITIVE ARTIFACTS IN AVIATION DISTRIBUTED COGNITION

Konstantinos Pechlivanis

Aviation is a complex and dynamic sociotechnical system. According to Harris (2013), there is a growing trend towards adopting a systems-based approach that recognizes the human-machine-task context as a collaborative cognitive system. The notion of distributed cognition recognizes that cognitive processes and activities extend beyond the boundaries of an individual's brain, encompassing interactions between people, tools, and the surrounding environment. Distributed cognition emphasizes how individuals, and their surroundings work together to accomplish mental tasks, sharing the cognitive load and enhancing problem-solving and decision-making.

Cognitive artifacts can be physical objects, like a calculator or a computer, digital tools, software, or even notational systems that aid in organizing and processing information. They play a crucial role in human cognition by augmenting and complementing our mental capacities, ultimately improving our overall cognitive performance and efficiency. In the aviation industry, cognitive artifacts (like EFBs, HUDs, aircraft health monitoring tools, and FMS, to name a few) have become essential components of daily operations. They reduce cognitive demands on aviation professionals and contribute to the overall safety and efficiency of air travel. By offloading routine tasks and providing access to critical information, these artifacts are indispensable in modern aviation, demonstrating the significance of human-technology collaboration in the industry (Ziakkas et al., 2023a). The aviation industry has witnessed a paradigm shift with the integration of AI applications into various operational aspects, which has significantly transformed the cognitive landscape of the aviation industry (Ziakkas et al., 2023b). AI systems are currently acknowledged as crucial cognitive artifacts that augment distributed cognition in aviation (Ziakkas et al., 2023a). The AI systems function as cognitive artifacts, increasing the dispersed cognition of pilots, air traffic controllers, and other professionals in the aviation industry. This article provides a limited overview of AI applications, highlighting current advancements in the field:

AI-based weather forecasting tools process vast amounts of meteorological data and provide real-time updates, enabling more accurate flight planning and decision-making. The development of **autonomous aircraft systems**, including drones and remotely piloted vehicles, exemplifies AI as a cognitive artifact.

AI-based predictive maintenance systems improve the overall efficiency of aviation operations by providing aircraft technicians with real-time information and diagnostics.

AI-assisted decision support systems analyze vast amounts of data, including air traffic, weather, and aircraft performance, to recommend route changes, or emergency procedures.

AI-based Natural Language Processing (NLP) applications can convert spoken language into text, facilitating more efficient and accurate communication between pilots and air traffic controllers. These tools reduce the risk of miscommunication and language-related errors, enhancing safety and efficiency in the air traffic management domain.

REFERENCES

Harris, Donald. (2013). Engineering Psychology and Cognitive Ergonomics. Applications and Services: 10th International Conference, EPCE 2013, Held as Part of HCI International 2013, Las Vegas, NV, USA, July 21-26, 2013, Proceedings, Part II. 10.1007/978-3-642-39354-9.

Ziakkas, D., Harris, D., Pechlivanis, K. (2023a). Towards eMCO/SiPO: A Human Factors Efficacy, Usability, and Safety Assessment for Direct Voice Input (DVI) Implementation in the Flight Deck. In Harris, D., Li, WC. (eds) *Engineering Psychology and Cognitive Ergonomics. HCII 2023*. Lecture Notes in Computer Science (Vol. 14018). Springer, Cham. https://doi.org/10.1007/978-3-031-35389-5_15

Ziakkas, D., Vink, L.-S., Pechlivanis, K., & Flores, A. (2023b). Implementation Guide for Artificial Intelligence in Aviation: A Human-Centric Guide for Practitioners and Organizations, HFHorizons, Oct 15 2023, independently published, ISBN 9798863704784.

2.6 THE VTR CASE STUDY

Evey Cormican

Training pilots by using VR changes everything. Numerous applications have adopted VR since the introduction of consumer-grade hardware. One of the primary benefits is the realistic depiction of the flight deck. This characteristic offers significant benefits in comparison to conventional flight simulators, such as decreased expenses associated with training, enhanced competence during aircraft familiarization exercises, and improved information retention. Visionary Training Resources (VTR) was established by pilots who possess considerable expertise in safety and training and recognize the criticality of efficacious training in the 21st century. The organization is likewise investigating additional applications of VR within the aviation sector, encompassing staff training, induction, soft skills development, and interviewing.

2.6.1 The Genesis of Visionary Training Resources (VTR)

VTR emerged from the vision of Evey Cormican, an airline captain who transitioned into entrepreneurship. Today, VTR boasts a team of 38 individuals, including senior executives and investors connected through the Kellogg network, marking a testament

Artificial Intelligence Training and Operations

to Evey's journey of resilience and innovation. Flight deck flows and procedures are the current focus of VTR. The objective is to guarantee a thorough comprehension of ground procedures starting from preflight to end-of-runway and from end-of-runway to the gate. Distractions and ATC integration happen in the next phase. VTR consults with customers about the decision-making process regarding educational resources and curriculum planning. VTR realizes it is important to carefully consider how the headset aligns with the learning objectives and overall curriculum structure, as well as the logistical considerations such as inventory management.

2.6.2 VTR Products Show Positive Results in Proof-of-Concept (POC)

VTR conducted a proof-of-concept study with JetBlue Airlines. The study was conducted with 70 new pilots who were placed into one of two groups. One group had access to all standard JetBlue study materials, and a second group had access to the exact same study materials, with the addition of a VR flight deck created by VTR. The study's findings revealed that pilot training had a variety of advantages, including substantially less instructor assistance during simulator sessions and shorter simulator session completion times. Furthermore, the VR group demonstrated considerable levels of interest, engagement, and viability as a training technology. Finally, pilots showed that they wanted more VR in their training, with 75% agreeing that including VR in all flows and procedures would give them sufficient training that they would be able to execute them without additional instruction. Furthermore, 78% of pilots agreed that if they had more flows and procedures in the headset, they would be able to pass their validations with greater ease. Overall, the study results were a success in demonstrating the value of VR for pilot training.

2.6.3 Incorporating VR into Pilot Training

VR headsets promote a blended learning approach that fits easily into the pilot training ecosystem and should be preferred over other solutions, in the areas of:

- Immersive Experience / Interactive Training / Scenario Replication / Cost-Effectiveness
- Flexibility and Accessibility: VR training can be conducted anywhere and at any time, providing flexibility for pilots to practice at their own pace and convenience. Additional self-paced iPad instruction is also being considered.

2.6.4 Generating and Using VR Data

The headsets collect significant amounts of data from the training modules. VTR uses this data to drive AI, ML, and advanced data analytics to fortify offerings, such as automation, non-normal procedures, or maneuvering tasks, that can diversify training modules and cater to various skill levels. All data are de-identified. VTR implements robust data management practices, including the ability to identify headsets for tracking and analysis purposes. VTR also ensures compliance with privacy regulations and addresses union concerns. Any data collected by VTR is made available to the airline for comprehensive training analysis and evaluation. Data tracking mechanisms are designed to associate usage with individual pilots or training sessions, facilitating targeted analysis and feedback.

2.6.5 THE FUTURE OF VIRTUAL REALITY AIRLINE TRAINING

With the integration of VR training, the curriculum could be enhanced to include more interactive and immersive learning experiences incorporating VR simulations for practicing complex procedures, emergency scenarios, and decision-making exercises. Additionally, the curriculum could emphasize collaborative training exercises in a multiplayer VR environment to enhance crew coordination and communication skills. Instructors could also leverage VR analytics and performance tracking tools to provide targeted feedback and support to individual pilots based on their training needs. Overall, the curriculum should be designed to maximize the benefits of VR technology while aligning with the airline's training objectives and regulatory requirements, including the FAA and EASA. The VTR five-year growth plan leverages the initial VR leadership position with expanding pilot training initiatives, including multi-player/crew resource management, non-normal emergency scenarios, remote viewing, air work scenarios, and recurrent training integration. VTR will invest deeply in R&D, create cutting-edge products, integrate with airlines and regulators, and expand relationships with communities, including youth outreach/aviation workforce development.

3 Artificial Intelligence in Traffic Management

3.1 APPLICATIONS OF AI IN TRAFFIC MANAGEMENT AND ITS CHALLENGES

Lea-Sophie Vink

Many regions worldwide are experiencing resistance to the increasing traffic congestion in urban areas. Primarily because of the congestion and pollution problems caused by excessive traffic on the highways. The road transportation sector is prone to disruption and can benefit from AI, like other transportation industries. At the core of the industry are the benefits that can be implemented into the individual vehicles themselves and the way that the networks and roads are managed to sustain them (Hasan et al., 2020). This may be a bit of an unanswerable question at the moment as the technological developments required for each are occurring in different spaces, but they will likely develop off each other and the network will self-adjust depending on what the market is inclined towards. For example, it is often stated that people's love of driving may slow down the transition to self-driving vehicles, but this may be offset by other needs such as moving to greener energy vehicles or disruptions to transportation methods for large goods, etc. Regardless, the implementation of AI can be looked at through the lens of these two areas and this is the subject of this chapter.

3.1.1 Individual Vehicles – Challenges and Benefits

There has been a wide range of academic research conducted not only into the concepts of self-driving and smart-vehicle technology. Papers such as those by Ryan (2020) cover the key ethical, legal, and social elements of what this technology can bring. Others such as Coicheci and Filip (2020) outline the technical challenges that still need to be overcome to implement seamless self-driving technology. But there can be no doubt, that significant advances have been made in the past decade in this area with sizable investments by some of the world's largest companies such as Google and Apple into solving the issues.

3.1.1.1 The Challenges for Autonomous Vehicles

The age of autonomous vehicles (AV) has been underway for some years now with many major cities publishing reports on their usage and legislators searching for ways to govern the ethics and liability issues concerning safety and responsibility. A synthesis of these reports tends to reveal similar themes and trends in the challenges for AVs namely:

3.1.1.1.1 Overall Traffic Management within Cities and Open Highways

This is generally the responsibility of city planners, police enforcement and legislators but reports suggest that the market has a key role to play here. For example, cars that can communicate with each other could travel more closely together and reduce congestion which can lead to shorter travel times, but this could have the opposite effect with commuters not stuck in traffic opting to take longer routes to enjoy their commute time. Additionally, external threats such as the massive increase in home office usage after the global pandemic suggest people might not actually want to be commuting much at all. This is coupled, at least in Europe with the increasing focus on pedestrianizing much of the old city centers. The market has a key role here in determining the speed at which AVs develop and become commercially viable (Milford et al., 2019).

3.1.1.1.2 Infrastructure

AVs use cameras and sensors to navigate and determine their place in space and time. Road infrastructure is required to be maintained at perhaps much greater rates than currently, and with more investment required, this could cost economies a great deal more. As mentioned above, there are social trends resisting the use of cars more and more, particularly in Europe, where other public transport methods such as cycling have seen a large renaissance. Increasingly, the public may not want to fund large scale infrastructure projects aimed at keeping cars on the road.

Another area requiring significant resilience and continuous service is the data-link and mobile phone technologies that will allow AVs to be able to communicate and navigate. Boer et al. (2017), for example, highlight the implementation of 5G networks and the resistance many nations faced when deploying them as a direct contributor to the slowing down of developing AV technology, especially in Europe. Curiously, people's trust in the reliability of cellular technology seems to be instinctively higher than the actual resilience of those networks. Ask any mariner what happens to GPS coverage in and around islands, or any driver stuck in a tunnel, and the vulnerability of the technology coverage to enable AVs is clear.

3.1.1.1.3 Legal, Ethics and Liability

As with other industries, individual human beings still hold licenses to drive vehicles. The issue of responsibility and liability in accidents has still not been solved. It will be tricky in the transitional years where both humans and machines drive cars side by side. Ryan (2020) covers many more of the ethical and social impacts of AVs but outlines that safety and prevention of harm need to continue to be at the forefront of all AI systems within individual vehicles. The key issue is how the AI systems will make decisions in the event of an accident. De Sio (2017) proposed a scenario where an AI-controlled vehicle, programmed to prioritize safety, may choose to swerve towards a cyclist wearing a helmet in a situation where it must decide between avoiding cyclists. This decision assumes that a cyclist wearing a helmet is more likely to survive a collision. However, this may result in individuals attempting to manipulate algorithms, such as fewer people wearing helmets when riding because they think it will prompt AV to drive more cautiously near them.

Artificial Intelligence in Traffic Management 51

3.1.1.1.4 Wider Social Impacts

Ryan (2020) cites a wide range of social impacts already observed but mentions that many may not have even been anticipated yet. The loss of the joy of driving is often cited as an existential threat posed by AVs. There appear to be gender differences in the trust given to AVs and concern about the inclusivity of Avs, although the potential for people who cannot drive to be included is a promising feature. But the unseen social impacts are still yet to be determined. How will travel behaviors truly change with the majority of the fleet of vehicles converted to self-driving? What happens if there is decreased urbanization? And of course, there are the looming environmental questions.

3.1.2 NETWORK AND TRAFFIC MANAGEMENT

At the macro-network control and traffic management level, when considering all of the different types of Avs, another story begins to emerge. Traffic management in cities and countries can be more closely linked to Air Traffic Management. AI technologies here pose a lot more benefit with less of the obstacles and issues surrounding individual AVs. The trends suggest that optimizing traffic management first could more easily pave the way for mixed vehicle fleets and ultimately a transition towards fully self-driving fleets.

3.1.2.1 The Concept of Operations

Today, traffic management consists of several key elements. Fundamentally, the network consists of roads that are navigated by drivers. Drivers choose routes based on time, distance, and perceived traffic congestion. Already common in most vehicles, on-board navigation systems increasingly rely on the internet to download the best routes, which are given by software algorithms based on data analysis. Google Maps, Apple Maps, or Waze are commonly used for this purpose, and the algorithms behind them rely on the data transmitted by thousands of users, each relaxing their positions, which allows the algorithms to optimize individual traffic around each other.

But vehicles are managed at various points along their route by traffic signaling, traffic lights, and occasionally manual input from, for example, road works or emergencies. Traffic signaling is usually controlled in large operations centers for cities and highways, and although traffic lights can be controlled here, many parts of the world still use traffic lights that are individually programmed in sequences (Ouallane et al., 2022).

3.1.2.2 Towards a more Integrated Solution

Increasingly, though, the solution for total traffic management systems is moving in the direction of a fully integrated system. Right now, most nations in Europe utilize fragmented management systems. For example, highway networks are controlled separately from city monitoring activities and traffic management within cities. The data sources are also disjointed. For example, highway monitoring systems like those that watch the M25 surround London rely upon close-circuit camera systems that belong to the organization but bring in traffic data from private companies like Google or Apple. There are cyber-security threats here as well, which are frequently cited (e.g., Ouallane et al., 2022).

However, a solution that brings together all elements of the grid would potentially provide major benefits. Routing is done individually but could be centralized via parameters at an AI and passed out through the platforms. The benefits to fuel consumption, CO_2 emissions, average waiting time, and average travel distance could all be reduced if these solutions are found. The idea would be that traffic signals, traffic lights, ad-hoc manual inputs, desired routing, and traffic flow and complexity data are all fed to an AI system that could optimize the usage of roads in real-time and provide routing solutions to thousands of vehicles (Ullah et al., 2019).

3.1.3 MERGING OF TWO DIRECTIONS

At the heart of the discussion on AI in road traffic management are two domains: individual AV and gross-network management. The connection between the two involves a modern kind of information and communication exchange called 'vehicle to everything (V2X)', where individual vehicles communicate directly with network management systems. It will rely on future communications networks not yet implemented (such as 6G technology) (Wang et al., 2019). Ultimately, AVs in the future could talk to each other, the road itself, network management, and cloud environments such as Google or Apple products, allowing a continuous smart optimization of routing. According to Boban et al. (2018), this will depend upon the rapid rollout of communications networks, but conceivably, the benefits of V2X communication may be the future direction of road management for all users. Where exactly the AI sits and who owns it remains to be seen, but the merging of these two domains is rapidly evolving.

REFERENCES

Boban, M., Kousaridas, A., Manolakis, K., Eichinger, J., & Xu, W. (2018). Connected roads of the future: use cases, requirements, and design considerations for vehicle-to-everything communications. *IEEE Vehicular Technology Magazine*, 13(3), 110–123.

Boer, A., van de Velde, R., & de Vries, M. (2017). Self-driving vehicles (SDVs) & geo-information. Retrieved from https://www.geonovum.nl/uploads/documents/Self-DrivingVehiclesReport.pdf.

Coicheci, S., & Filip, I. (2020, May). Self-driving vehicles: current status of development and technical challenges to overcome. In *2020 IEEE 14th International Symposium on Applied Computational Intelligence and Informatics (SACI)* (pp. 000255–000260). IEEE. Doi: 10.1109/SACI49304.2020.9118809

De Sio, F. S. (2017). Killing by autonomous vehicles and the legal doctrine of necessity. *Ethical Theory and Moral Practice*, 20(2), 411–429.

Hasan, M., Mohan, S., Shimizu, T., & Lu, H. (2020). Securing vehicle-to-everything (V2X) communication platforms. *IEEE Transactions on Intelligent Vehicles*, 5(4), 693–713.

Milford, M., Anthony, S., & Scheirer, W. (2019). Self-driving vehicles: key technical challenges and progress off the road. *IEEE Potentials*, 39(1), 37–45.

Ouallane, A. A., Bahnasse, A., Bakali, A., & Talea, M. (2022). Overview of road traffic management solutions based on IoT and AI. *Procedia Computer Science*, 198, 518–523.

Ryan, M. (2020). The future of transportation: ethical, legal, social and economic impacts of self-driving vehicles in the year 2025. *Science and Engineering Ethics*, 26(3), 1185–1208.

Ullah, H., Nair, N. G., Moore, A., Nugent, C., Muschamp, P., & Cuevas, M. (2019). 5G communication: an overview of vehicle-to-everything, drones, and healthcare use-cases. *IEEE Access*, 7, 37251–37268.

Wang, J., Shao, Y., Ge, Y., & Yu, R. (2019). A survey of vehicle to everything (V2X) testing. *Sensors*, 19(2), 334.

3.2 TRAFFIC PREDICTION AND OPTIMIZATION
Anastasios Plioutsias and Dimitrios Ziakkas

The aviation industry has undergone a significant transformation with the introduction of artificial intelligence (AI). AI has revolutionized aviation traffic management, drastically changing flight path planning, schedule optimization, and overall airport operations. The role of AI in modern aviation is important for optimizing traffic prediction and management. Traffic prediction in aviation involves forecasting the number and movement of aircraft within a specific airspace or approaching/departing an airport. This predictive capability is fundamental for planning, from air traffic control to scheduling and ground operations. AI algorithms leverage extensive datasets, including historical flight data, weather information, and real-time airspace conditions, to accurately predict traffic patterns. Machine learning models and intense learning networks are at the forefront, capable of identifying complex patterns and making predictions based on multifaceted criteria (Li et al., 2019).

3.2.1 Traffic Optimization

Traffic optimization involves predicting patterns and finding ways to improve them. This means optimizing flight routes to avoid congestion, scheduling flights to make the most of airport and airspace capacity, and making airport ground operations more streamlined. Moreover, AI algorithms can suggest more efficient routes based on weather conditions, airspace restrictions, and predicted traffic flows. AI can help airlines and airports optimize their schedules by predicting peak travel times and potential bottlenecks, reducing delays, and improving the passenger experience. Furthermore, AI can also optimize the allocation of gates, baggage handling, and other ground services to ensure smooth operations. The benefits of traffic optimization are significant, ranging from reduced fuel consumption and emissions to improved punctuality and passenger satisfaction. AI's impact on traffic prediction and optimization in aviation is significant, offering a glimpse into a future where flights are safer, more efficient, and less prone to delays. As technology advances, the integration of AI in aviation will undoubtedly deepen, driving further improvements in traffic management and operational efficiency.

3.2.2 Future Directions in AI for Aviation Traffic Management

As AI algorithms advance, they will better predict traffic patterns accurately and over longer periods of time. To accomplish this, it will be necessary to enhance current machine learning models and integrate more recent AI methodologies, including generative adversarial networks (GANs) and reinforcement learning. These developments will enable more adaptable and dynamic traffic management strategies to accommodate unforeseen circumstances, like sudden weather changes or operational disruptions (Vlahogianni et al., 2014). More diverse and real-time data sources, such as satellite imagery, social media trends (predicting demand), and Internet of Things sensors across airports and aircraft, will be integrated to enhance the data pool from which AI models can draw. This will lead to more precise predictions and optimizations, facilitating a more responsive and resilient aviation ecosystem. We can expect the emergence of collaborative AI systems where multiple models share information and learn from each other (Osaba et al., 2017). As autonomous aircraft and commercial drones become more common, AI will play a crucial role in integrating these new entrants into the airspace without disrupting existing traffic flows. Achieving this will require innovative traffic management solutions that can dynamically adapt to a more complex and crowded airspace (Polson & Sokolov, 2017). Future AI developments will likely focus on maximizing these benefits, aligning with global efforts to combat climate change.

3.2.3 SAFETY AND SECURITY

The safety and security of AI systems are of utmost importance. They should be designed to enhance, rather than compromise, operational integrity and passenger safety in aviation. AI systems should adhere to rigorous safety and security standards in designing, testing, and implementing them to achieve this. Regular updates and patches should also be applied to address new vulnerabilities. The successful integration of AI into aviation traffic management requires a comprehensive approach that considers privacy, accountability, transparency, equity, safety, and security concerns. As AI technologies continue to advance, ethical guidelines and regulatory standards must also evolve to ensure that the implementation of AI in aviation traffic management benefits all stakeholders responsibly and equitably (Zhang & Haghani, 2015).

3.2.4 TECHNOLOGICAL CHALLENGES IN AI FOR AVIATION TRAFFIC MANAGEMENT

High-quality data and seamless integration are essential for the effective deployment of AI systems. The efficacy of AI relies significantly on the quality, precision, and thoroughness of the data utilized for training and decision-making. One of the challenges in this regard is dealing with inconsistent, incomplete, or inaccurate data, which can cause flawed AI predictions or optimizations. To overcome this challenge, it is necessary to establish robust data management frameworks and standards to ensure data integrity and reliability.

Real-time processing capabilities are also crucial in aviation traffic management, especially in dynamic and rapidly changing environments. Furthermore, integrating AI systems across different platforms, jurisdictions, and technologies in aviation requires high interoperability and standardization. However, diverse systems and standards can hinder the seamless integration of AI solutions. Therefore, it is necessary to work towards global standards and protocols for AI in aviation to ensure compatibility and interoperability. As AI systems become essential to aviation traffic management, they become vulnerable to cybersecurity threats. Protecting these systems from attacks is fundamental. Implementing state-of-the-art cybersecurity measures, continuous monitoring, and adopting a proactive stance on cybersecurity can ensure that AI systems are secure against increasingly sophisticated cyber threats. Building trust in AI systems among air traffic controllers (ATCO), pilots, and passengers is challenging. Engaging with regulatory bodies early in the development process and contributing to developing AI-specific regulations and standards can help overcome this challenge.

In conclusion, addressing the technological challenges is crucial for successfully integrating AI into aviation traffic management. Collaboration among technology developers, aviation authorities, and regulatory bodies can pave the way for safer, more efficient, and more sustainable aviation operations.

REFERENCES

Li, L., Lv, Y., & Wang, F. Y. (2019). Traffic signal timing via deep reinforcement learning. *IEEE/CAA Journal of Automatica Sinica*, 6(3), 703–713.

Osaba, E., Yang, X. S., Diaz, F., Onieva, E., Masegosa, A. D., & Perallos, A. (2017). A review on artificial intelligence in traffic management. *Transport*, 33(3), 711–725.

Polson, N. G., & Sokolov, V. O. (2017). Deep learning for short-term traffic flow prediction. *Transportation Research Part C: Emerging Technologies*, 79, 1–17.

Vlahogianni, E. I., Karlaftis, M. G., & Golias, J. C. (2014). Short-term traffic forecasting: where we are and where we're going. *Transportation Research Part C: Emerging Technologies*, 43, 3–19.

Zhang, Y., & Haghani, A. (2015). A gradient boosting method to improve travel time prediction. *Transportation Research Part C: Emerging Technologies*, 58, 308–324.

3.3 ROLE OF AI IN UNMANNED AIRCRAFT SYSTEMS AND THE DRONE TRAFFIC MANAGEMENT SYSTEM

Andrew Black

The future of aviation is in rapid advance to unify Unmanned Aerial Systems (UAS) as part of national aerospace activity. This is evidenced by the growing weight of research into areas such as airspace integration (Thipphavong et al., 2018) and UAS design (Sadraey, 2020). There is however a void of academic information relating to Human Factors (HF) within the field, as discussed by Black et al. (2023). Indeed, Sadraey's (2020) taxonomy for design groups fails to overtly identify HF as a design discipline.

Such omissions create a pre-condition for poor UAS design: **unmanned does not mean *unhumanned*.** Wherever there are humans, there are HF and *that* poses a problem if they remain unaccounted for in a world that relies on increasing automation complexity. This chapter will discuss how Machine Assistance for Remote Aviation (MARA) might help augment HF for Visual Line of Sight (VLOS) operations and as a result, improve safety.

3.3.1 Augmenting the Envelope: Machine Assistance to Remote Aviation (MARA)

The term *"Remote Aviation"* has been employed in this case to acknowledge that while UAS refers to an entire system, its component parts on occasion must be treated separately, be that for research, adaptation, or update. For example: alerting systems, transmitter interfaces, training programs, remote pilots (RPs), or the vehicle itself. All are constituent parts of an UAS, but the entire UAS may not need review. Rather, when viewed as separate elements, they relate to remote aviation in general as opposed to one specific completed system. For example, the term UAS refers solely to the machine (including the powerplant, engines and C2). It needs a RP to operate it, who will have some degree of training. Once these elements are brought together, it becomes a UAS.

To understand how the system fits together and might be improved, these elements must necessarily be treated separately, while respecting their relationship to remote aviation as opposed to manned. Of course, to close the loop, any change to that element still requires an assessment of the UAS as a whole. Too many accidents have been caused by the failure to assess incremental component "improvement" without assessing how the wider system is affected by the change. Furthermore, the term *"remote aviation"* also acknowledges that humans – to some degree – are involved. We have the term *Remote Pilot* to define the operator of an *unmanned* system: a tangible example of unmanned but not unmanned. Surely this is an obsolete term in the era of aviation HF?

It is therefore posited that the use of the word *unmanned* should be reviewed. Without doubt, the use thereof is important in expanded definitions, but perhaps *"remote"* better reflects the human involvement from concept to crash. UAS versus *Remote* Aerial System instantly frames different mental models: the first without humans, the second with. It is argued that such psychological framing is critical to flight safety.

3.3.2 What Is Machine Assistance?

Literature shows that the accepted definition of machine assistance requires machine learning (input), a problem (process 1), a human user (process 2), and a machine generated resolution (output). This output is not *necessarily* automatic, instead presenting a decision option to the system user. Humans are again an integral part of an apparently automated system.

Human programmers are required to generate the code for machine learning and consequently may introduce unconscious bias. One recent example with Google's Nest AI camera had some very socially unacceptable results: people of color were mistaken for animals (The New York Times, 2023). This article is not about the problems with machine learning, though, but it is useful to caveat any benefit by reminding ourselves of significant failure.

This caveat highlights how important it is for humans to be in-the-loop (ITL) with decision-making processes. For our single-crewed, VLOS RP, visual modality conflict is a safety-critical problem (Wickens et al., 2004; Black et al., 2023).

However, the scientific proximity to an Out-of-the-Loop (OOTL) condition for these RPs remains an area for research. Qualitative data from Black et al.'s (2023) study would suggest VLOS is an area where machine assistance would be beneficial.

The existing digital infrastructure available to manned aviation may present a path of least resistance for expedited implementation of MARA. For example, Mode-S transponders allow ATCO to see what automation selections have been made by pilots (typically: heading, speed and altitude). The same technology could be employed to transmit safety critical signals to RPs.

3.3.3 MARA AND EASA

EASA is framing how AI can support all types of aviation and has produced three classifications of AI (EASA, 2023). The discussion so far relates to Level 1 and arguably Level 2, but how useful are the classifications in Level 2? *Cooperation* refers to situations where entities work together in support of achieving individual goals, but under *collaboration*, those same entities work towards a shared goal. While human-AI teaming is an appropriate suggestion for Level 2, it is posited that the delineation between *collaboration* and *cooperation* lacks sufficient granularity. However, if the intent of cooperation is for both machines and humans to learn from each other, then perhaps there is validity to those descriptors.

3.3.4 FUTURE RESEARCH

In order to truly identify how machine assistance can help VLOS RPs, more research into RP HF *has* to be conducted. It is argued that incentives to develop areas such as urban air mobility put too great an emphasis on vehicle technology at the expense of human interaction with the system. The true extent of OOTL and ITL remains unknown for VLOS operations and may well be an area that MARA could support. Evidence suggests that alerting systems could be greatly improved, drawing on lessons learned from manned aviation (Black et al., 2023), once again an area where MARA may be of use. Existing digital aviation services may provide an expeditious pathway for MARA research. Finally, EASA's Level 2 AI classification might perhaps benefit from scientific scrutiny.

3.3.5 CONCLUSION

To reiterate a theme: **unmanned does not mean *unhumanned***; wherever there are humans, there are HF. By framing drone aviation through the term *unmanned aerial system*, flight safety is at risk; it creates a pre-condition for OOTL behavior. A subtle change to *remote* may mitigate that mindset. From concept to crash, HF *must* become as integral to remote aviation as they are to manned. VLOS RPs might be assisted by the appropriate use of machine assistance, but for that to happen, more directed research is required. MARA concepts may complement ongoing research into Drone Traffic Management Systems (also known as Unmanned Traffic Management [UTM]). Finally, it is argued that by continuing to use the term *unmanned*, remote aviation distances itself from any benefit HF principles might offer.

REFERENCES

Black, A., Scott, S., & Huddlestone, J. (2023). Attitude adjustment: enhanced ATTI mode for remote pilots. *Engineering Psychology and Cognitive Ergonomics*, 3–17. https://doi.org/10.1007/978-3-031-35389-5_1

EASA. (2023, May). EASA artificial intelligence roadmap 2.0 – A human-centric approach to AI in aviation | EASA. https://www.easa.europa.eu/en/document-library/general-publications/easa-artificial-intelligence-roadmap-20

The New York Times. (2023, May 22). Google's photo app still can't find gorillas. And neither can Apple's. *The New York Times* – Breaking News, US News, World News and Videos. https://www.nytimes.com/2023/05/22/technology/ai-photo-labels-google-apple.html

Sadraey, M. H. (2020). *Design of Unmanned Aerial Systems*. John Wiley & Sons.

Thipphavong, D. P., Apaza, R., Barmore, B., Battiste, V., Burian, B., Dao, Q., Feary, M., Go, S., Goodrich, K. H., Homola, J., Idris, H. R., Kopardekar, P. H., Lachter, J. B., Neogi, N. A., Ng, H. K., Oseguera-Lohr, R. M., Patterson, M. D., & Verma, S. A. (2018). Urban air mobility airspace integration concepts and considerations. In 2018 *Aviation Technology, Integration, and Operations Conference*. https://doi.org/10.2514/6.2018-3676.

Wickens, C. D., Gordon, S. E., & Liu, Y. (2004). *An Introduction to Human Factors Engineering*. Prentice Hall.

3.4 ROLE OF AI IN UNMANNED AIRCRAFT SYSTEMS AND THE DRONE TRAFFIC MANAGEMENT SYSTEM

Andrew Black

There are several synonyms for Drone Traffic Management Systems; however, the Federal Aviation Administration provides a description for UTM, which appears to be universal. It is defined as airspace for Beyond Visual Line of Sight vehicles, below 400 feet above ground level, where air traffic services are not provided. Such uncontrolled airspace in theory sounds like an ideal place for remote operations, however, at present, there are a plethora of issues, not least of which is the risk of collision with other vehicles and buildings. Machine Assistance to Remote Aviation (MARA) and AI will become key tenets to help humans monitor these ever more automated environments. It should always be remembered though: removing humans does not *necessarily* improve safety.

3.4.1 THE WORLD BELOW 400 FEET ABOVE GROUND LEVEL

One must only look out of the window to see an obvious problem: collision risk. With buildings, vehicles, people, terrain and of course, other drones. At present, this is uncontrolled airspace and there are good reasons for that: it is difficult for humans to decipher. Primary radar (simple scan, returns all hits) would generate too much ground clutter for distinguishing drones and obstacles from one-another. Secondary Surveillance Radar, on the other hand, relies on aircraft fitted with suitable transponders that display specific target information to a controller. Neither system is effective at low levels with multiple returns and no method of human control. The contemporary equipment works well for manned aviation, at altitude or in a controlled environment around an aerodrome. Just as manned aviation has evolved

Artificial Intelligence in Traffic Management

to cope, so remote aviation is doing the same. An apt term has been developed to describe multiple drones operating in close proximity: swarming (Devos et al., 2018).

3.4.2 Towards Control in the Uncontrolled

In manned aviation, there one of the most basic life-saving concepts: See-and-Avoid. As a pilot, you look for danger, be it an obstacle or an aircraft, and once spotted you avoid it. This concept formed the basis for drone Detect-and-Avoid technology. As basic mitigation, just like manned aviation, the technology requires either a dedicated corridor (Aldao et al., 2022) or an onboard intelligent vision system (Devos et al., 2018). However, significant advances in swarming research appear to be addressing the issue. In keeping with the best history of automation irony, the *uncontrolled* airspace is becoming controlled by automation. (Ross et al., 2021; White, 2024). However, even in this highly automated scenario, human input is still required. For example, humans still prepare flight plans, prepare drones, and monitor systems. It is this last point where MARA and AI might really create synergy because monitoring is, in fact, the hardest task the human brain can perform (UK CAA, 2013). At a primeval level, we are still hunters and protectors, so to sit passively is simply unhuman.

3.4.3 MARA Monitoring

Monitoring is an important part of situation awareness (SA) by aiding at notice, understand and think ahead levels. However, to be effective, specific and appropriate parameters must be defined. At present, it appears this role is undefined, and yet in manned aviation, there is a safety-critical emphasis on the same role (in this context, that would be an ATCO).

MARA could be employed to support Level 1 (Notice) and Level 2 (Understand) SA, leaving Level 3 (Think Ahead) to be performed by our human monitor: the real expert. With two humans, this would be known as over-lapping SA and helps generate spare mental capacity for the final decision-maker. Machines process all the data from the swarm and are programmed with data such as proximity limits, speed limits, or power limits, for example. Once a problem arises, a suggested course of action is suggested to our human-ITL. Safety is not only maintained, but also enhanced by harnessing the intangible human benefits of experience, knowledge, and judgement. It may be the case that our MARA solution is accepted, but like humans, machines are not infallible. This hybrid solution would appear to bring the best of both worlds together for the monitoring task; however, more research is required.

3.4.4 Conclusions

This chapter analyzes the problems faced by UTM in sub-400-feet airspace. While levels of drone automation have necessarily increased to facilitate such operations, how the human role has changed is not being given sufficient credence: monitoring is the hardest and most safety-critical of aviation functions. That is where MARA can help by supporting SA principles. The extra capacity afforded to our humans can then be employed with other related safety-critical tasks. It is imperative that future research include the human-machine-interaction in its entirety; at present, it feels like humans are being forgotten. As stated in a previous chapter, unmanned does not mean *unhumanned*. Perhaps it's time a term like *Remote Aviation* was adopted to better reflect future complexity.

REFERENCES

Aldao, E., González-de Santos, L., & González-Jorge, H. (2022). LiDAR based detect and avoid system for UAV navigation in UAM corridors. *Drones*, 6(8), 185. https://doi.org/10.3390/drones6080185.

A. Devos, E. Ebeid and P. Manoonpong, "Development of Autonomous Drones for Adaptive Obstacle Avoidance in Real World Environments," *2018 21st Euromicro Conference on Digital System Design (DSD)*, Prague, Czech Republic, 2018, pp. 707-710, doi: 10.1109/DSD.2018.00009.

Ross, A. L., Rozell, N., Allamraju, R., & Jacob, J. D. (n.d.). Evaluation of UAS Swarming in a BVLOS Environment. In *AIAA AVIATION 2021 FORUM*. https://doi.org/10.2514/6.2021-2336

UK CAA. (2013). CAA paper 2013/02: Monitoring matters – Guidance on the development of pilot monitoring skills. UK Civil Aviation Authority. https://publicapps.caa.co.uk/modalapplication.aspx?appid=11&mode=detail&id=5447.

White, J. T. (2024). Autonomous Multi-layer Integrated Macro-/Micro-Swarming Networked System-of-Systems for UAM. In *AIAA SCITECH 2024 Forum* (1–0). American Institute of Aeronautics and Astronautics. https://doi.org/10.2514/6.2024-1288

3.5 THE ROLE OF AI IN TRAINING OF TRAFFIC CONTROLLERS – MANAGERS

Lea-Sophie Vink

Like all other domains in Aviation, AI stands to benefit the training of ATCO in both predictable and uncertain ways. Across many parts of the world, ATCO training is increasingly under scrutiny, both for the length of time it takes to train an ATCO (about 3 years on average) and the decreasing number of new applicants, which is causing some organizations to become increasingly desperate as their work forces are aging and retiring without relief. With that in mind, it is crucial we continue to train new ATCOs but seek out ways to make use of AI to reduce some of the challenges in qualifying them. This chapter addresses several key opportunities AI brings to the training of ATCOs.

3.5.1 Overview of the Typical ATCO Training Cycle

In Europe, all ATCO must demonstrate the same basic competencies outlined and regulated by the European Aviation Safety Authority (EASA 2019/023/R) to be awarded a license. Typically, ATCOs follow the three phases outlined by ICAO: qualification and basic training in phase 1, unit-specific training including on-the-job training in phase 2, and then continuation and refresher training in phase 3. The basic training is usually about 6–9 months long, consisting mostly of classroom-based theoretical training with some simulator courses. Following basic training, ATCO trainees move onto concentrated simulator training for the given ATCO operating environment they are selected for (e.g., Tower controller or Area Controller). This is then followed by early on-the-job training at a unit. ATCO trainees will then go back and forth between units and simulators, building experience and passing exams for relevant competencies such as bad-weather scenarios, night scenarios, emergencies, etc.

3.5.2 Where AI Can Be Used to Improve and Speed Up Training

There are several key areas where AI can be used to help improve the training cycle. They generally fall into two key areas: AI within the simulators and selection of scenarios, and AI to assist and tailor individual solutions.

3.5.2.1 AI-Assisted Simulation

Simulation training is the main cornerstone of ATCO education. AI can significantly enhance these simulations by providing dynamic and adaptive scenarios that can teach and expose students to the things they need to practice most, providing real-time feedback that is unbiased and far more accurate than the instructors and examiners who invariably introduce their own biases (Abdillah et al., 2024). Machine learning can be used to understand how individual students are learning, what their strengths and weaknesses may be, and then pose realistic scenarios for them to deal with. The AI would also be able to help instructors develop more realistic scenarios. This can also enhance the teaching and help instructors prepare better for addressing the competencies most needed.

3.5.2.2 Personalized Learning Paths and Workload Management

Every ATCO trainee has unique strengths and weaknesses when learning and studying. AI can help analyze individual differences and tailor training programs specific to individuals. By providing these pathways, it can help trainees learn much faster and more accurately than ever before. Adding tools like virtual or augmented reality that are adapted to individual preferences and needs will help to enhance these pathways and the learning mechanisms (Zuluaga-Gomez et al., 2023). This will have the added benefit of improving the workload management of ATCO trainees because they will be better suited to their studying needs to meet their goals. For already qualified ATCOs needing specific refreshers, their profiles could be loaded from their initial training days and added to occurrence reports that recommend certain key points required for them to practice or maintain.

3.5.2.3 Performance Analysis and Feedback

AI-driven analytics can offer comprehensive insights into trainee performance. By tracking and analyzing various parameters, such as response times, decision accuracy, and communication effectiveness, AI can provide objective feedback. This data-driven approach enables trainers to identify specific areas for improvement and offers a basis for constructive feedback, fostering continuous improvement.

3.5.3 Conclusions

The integration of AI into ATCO training holds immense potential for revolutionizing how we prepare the next generation of ATCOs (Gosling, 1987). By utilizing AI for adaptive simulations, personalized learning paths, and cognitive load management, we can create a training environment that is not only more effective but also reflective of the dynamic and complex nature of real-world air traffic control.

REFERENCES

Abdillah, Risya & Moenaf, Henry & Rasyid, Luthfi & Achmad, Said & Sutoyo, Rhio. (2024). Implementation of Artificial Intelligence on Air Traffic Control - A Systematic Literature Review. 1-7. 10.1109/IMCOM60618.2024.10418350.

Gosling, G. D. (1987). Identification of artificial intelligence applications in air traffic control. *Transportation Research Part A: General*, 21(1), 27–38.

Zuluaga-Gomez, J., Prasad, A., Nigmatulina, I., Motlicek, P., & Kleinert, M. (2023). A virtual simulation-pilot agent for training of air traffic controllers. *Aerospace*, 10(5), 490.

3.6 MONITORING HUMAN PERFORMANCE, SAFETY INTELLIGENCE, AND FATIGUE RISK MANAGEMENT IN TRAFFIC ECOSYSTEMS

Spyridon Chazapis

3.6.1 Fatigue Risk and Workload Management Applications of Artificial Intelligence

Current approaches to fatigue risk management are characterized by a progressive evolution away from traditional prescriptive approaches toward Safety Management System principles. These Fatigue Risk Management Systems are *"data-driven means of continuously monitoring and managing fatigue-related safety risks, based upon scientific principles and knowledge as well as operational experience that aims to ensure relevant personnel are performing at adequate levels of alertness"* (ICAO, 2015).

Hazard identification is affected using a combination of predictive, proactive, and reactive methods: These include the analysis of fatigue reports submitted by operators, as well as data derived from self-report instruments (usually in the form of mobile applications) that assess alertness levels at various points during the flight. Reactively, data obtained from occurrence reports can provide data on fatigue as a contributory cause whereas an additional amount of data can be accumulated during the proactive activities of the Fatigue Risk Management Systems (examining emerging risks for which no prior data exists to facilitate a predictive analysis). Furthermore, biomathematical modeling of fatigue levels can be used to predict the impact of specific work/rest patterns – a useful tool for hazard identification at the scheduling level. AI can be used to overcome two significant obstacles associated with these methods:

- A significant amount of data is amassed, and statistical analysis can be difficult and time-consuming. Furthermore, it requires significant human involvement since this is an essentially manual process of identifying threats, assigning an appropriate level of significance, translating these

Artificial Intelligence in Traffic Management

threats into meaningful and measurable Safety Performance Indicators, and, eventually, instituting mitigation measures.
- A significant amount of the data collected is based on self-reporting. For a variety of reasons (Sieberichs & Kluge, 2021), operators may decide not to report incidents of fatigue or fail to accurately assess the associated level.

For the first item, AI can be applied in analyzing the collected data to reveal underlying patterns and potential threats that would be difficult to identify using traditional statistical methods. This analysis can be conducted using:

3.6.1.1 Machine Learning/Deep Learning/Natural Language Processing

The potential exists for using AI to provide a real-time, quantitative assessment of operator fatigue levels through the analysis of physiological indicators. These can include facial expressions using non-intrusive methods like infrared video cameras, the measurement of biometric data such as heart and respiration rates or galvanic skin responses using wearable devices (smart watches or rings), and the analysis of speech and voice data for atypical patterns that could indicate impending fatigue. In addition to these, fatigue levels can also be assessed indirectly by comparing cognitive performance to a baseline using methods like eye tracking or event response times. These methods can provide the possibility of real-time interventions during the early onset of fatigue, a capability that does not exist yet and offers the potential for a significant increase in operational safety (Lyons & Barber, 2019). For this reason, they are the topic of intense research and development efforts by various institutions (see the HIPNOSIS project or Blueskeye AI for two examples of such research). The use of AI for managing fatigue risks offers several significant advantages over the use of traditional statistical methods. Some of these are:

- **Scalability**
- **Reduction of the time required for data analysis**
- **Reduction of operational costs**
- **Data-driven insights**

In conclusion, the use of AI in fatigue risk management can address the need for analyzing large data sets through the use of easily scalable computing power, provide insights that are difficult to obtain through classic statistical methods, and enable new avenues for the identification of fatigue levels through methods that are impervious to the shortcomings of self-reporting.

REFERENCES

IATA/ICAO/IFALPA Fatigue Management Guide for Airline Operators, 2nd Edition (2015), International Civil Aviation Organization, 999 Robert-Bourassa Boulevard, Montréal, Quebec, Canada

Lyons & Barber. (2019). Driving innovation: a review of the latest trends in connected and autonomous vehicles. *Journal of Transport & Geography*, 77, 112–123.

Sieberichs, S., & Kluge, A. (2021). Why commercial pilots voluntarily report self-inflicted incidents: a qualitative study with aviation safety experts. *Aviation Psychology and Applied Human Factors*, 11(2), 98–111.

3.7 SAFETY ASSURANCE, SAFETY MANAGEMENT, AND THE DESIGN OF AIRSPACE/TRAFFIC MANAGEMENT
Lea-Sophie Vink

One of the most exciting opportunities that AI presents is in the realm of safety intelligence and the analysis of safety events. For many decades, experts have bemoaned the fact that we cannot ever technically know if we are 'safe right now.' Instead, we rely on the knowledge that, according to all data sources, we were safe 1 minute ago (US DoD, 2022). In the last five years, with the publication of strategies like Eurocontrol's 'weak-signal analysis' techniques, this has changed rapidly. Weak-signal analysis hunts for tiny bits of data that, when taken together, can give you a much more objectively informed picture. This chapter discusses how AI can take weak-signal analysis much further and provide an informed analysis for safety.

3.7.1 Human Performance and Fatigue as Examples of Weak Signal Analysis

Using the example of human performance and fatigue as areas where weak-signal analysis can be utilized, consider that if human performance decreases due to things like fatigue, high workload and stress, this can contribute to human error and reduced system efficiencies. In air traffic management, for example, the ATCO get more tired throughout their shifts. In effect, weak-signal analysis and using AI and machine learning to learn from this weak-signal analysis can hold a mirror to the way people imagine they are working. Consider the following case study taken from real data:

3.7.2 Why AI Will Take This and Make It Better, but with Challenges

The above case study shows how AI-driven tools, like Skyroster, facilitate such analysis without compromising individual privacy. This approach, contrasting with traditional top-down human analysis, advocates for a bottom-up methodology that focuses on understanding normal operations and identifying deviations. The selected case study also presents a proposed method for analyzing human contributions to occurrences, underscoring the importance of maintaining anonymity in data sets to ensure discussions on safety remain focused on operational rather than individual performance. Moreover, it critiques the conventional focus of organizations on managing and mitigating fatigue risk, suggesting a more comprehensive approach that includes measuring and assessing current conditions, modeling, and analysis, as exemplified by the FAA's fatigue risk management process (FAA, 2015).

3.7.3 How Does Safety Intelligence Feed Safety Assurance?

Using the weak-signal analysis allows for an objective view of safety performance. When preparing the safety assurance arguments in change management and implementation, as is required by most national legislation, this kind of knowledge can be used to form the basis for answering the question, 'What does today's operation look like?'

Any safety assurance argument normally hinges on whether a new change can demonstrate through verification of requirements and validation of the delivered changes if the change will continue to perform at today's standards, or ideally better (because why change in the first place?). If the objective safety intelligence picture is known, then this can form the basis of the quantification quite easily and safety objectives can be drawn up against this. That way any simulation, implementation or validation measures used to monitor changes can be held up against this measurement.

3.7.4 Next Steps Towards Parameterization

The article explored the potential of expanding AI safety intelligence systems beyond their current applications, suggesting that safety parameters could be integrated into larger, central processing units to enhance operations in sectors like power, train networks, and air traffic management (Salmon et al., 2012). While larger operations might benefit from a centralized approach, smaller ones, such as individual aircraft or ships, could adopt different strategies. The discussion points towards the development of a comprehensive concept by a European Air Navigation Service provider, aiming to utilize AI tools for collecting and analyzing safety intelligence from various sources, including ADS-B transmissions. This approach, focusing on weak-signal analysis, is highlighted as a significant advancement in addressing safety concerns, demonstrating the importance of asking the right questions to effectively utilize the data available. The narrative sets the stage for further discussion on implementing AI-driven safety systems that can adapt to the scale and specifics of different operations.

REFERENCES

FAA. (2015). Safety Management Systems for Aviation Service Providers (AC 120-92B). https://www.faa.gov/regulations_policies/advisory_circulars/index.cfm/go/document. information/documentid/1026670

ICAO. (2018). *Safety Management Manual (Doc 9859)* (4th ed.).https://www.icao.int/safety/safetymanagement/pages/guidancematerial.aspx

Salmon, P., Cornelissen, M., & Trotter, M. (2012). Systems-based accident analysis methods: a comparison of Accimap, HFACS, and STAMP. *Safety Science*, 50, 1158–1170. https://doi.org/10.1016/j.ssci.2011.11.009

US Do D. (2022). Human Factors Analysis and Classification System 8.0.https://www.safety.af.mil/Divisions/Human-Performance-Division/HFACS/

3.8 THE AUSTRO CONTROL CASE STUDY

Lea-Sophie Vink

In Chapter 1, we outlined that due to legal restrictions that require humans to remain in the loop and hold the liability so long as they hold a license to be a pilot, ATCO, maritime navigator, or train driver, the removal of humans from the loop remains, at least for the foreseeable future, a key element of any operational system. At the moment, most operations utilize some sort of central control over all of the technologies. But how much 'control' does a human really have? Increased automation levels can reduce human performance (for example, workload) and increase human error, which can lead to safety and performance issues. Our systems are now so sophisticated that the line between man and machine is quite blurred, as again in the example of the 737-Max accidents in Nigeria and Indonesia.

In our transportation systems, we now have an increasing conundrum:

1. How to keep humans in the loop, legally liable, responsible and able to intervene if,
2. The systems are increasingly automated, which can reduce human performance, increase human error, and decrease the ability of humans to make informed decisions.

One way to find the answer might be to change our ideas about what it means to be human in the first place. If we imagine a system where people are important and their performance is important to the system, but where technology controls that performance, considering things like stress, fatigue, and not being able to see what's going on around them, then the system could change the performance of people based on its needs. We just need to look at it the other way around to see what adaptive technology can really do. For the best way to sum this up, picture the plane flying the pilots instead of the pilots flying the plane. This chapter discusses a solution currently in development at Austro Control, Austria's Air Navigation Service provider. We can gradually implement increasingly automated technologies that allow individuals to maintain control, take responsibility, and stay involved by altering our approach to human performance and its supervision.

3.8.1 THE COMPONENTS AND MASTER SUPERVISORY DECISION-SUPPORT SYSTEM

Figure 3.8.1 displays the complete system overview. Austro Control has been studying numerous components for years. The notion is that operations room supervisors must make important decisions to control traffic. If traffic is heavy, supervisors have two options to relieve ATCO. They may:

1. Control the flow of traffic by implementing flow-control measures further down the network. This might involve notifying neighboring countries, or the European Network Control manager in Brussels, Belgium.
2. If flow measures are not enough, supervisors can open and close workstations depending on the number of staff at their disposal.

Supervisors have only two possibilities. They must balance personnel performance with procedures and technology controls. Supervisors need a tool to help them choose solutions. To manage staff best, supervisors need to know who is available and when based on their rostering and fatigue risk management parameters. Therefore, it was conceived that the tool should focus on showing the availability of staff. The tool takes objective system parameters from Task Complexity indices, human performance indices and weak-signals analysis (outline in the previous chapter)

Artificial Intelligence in Traffic Management

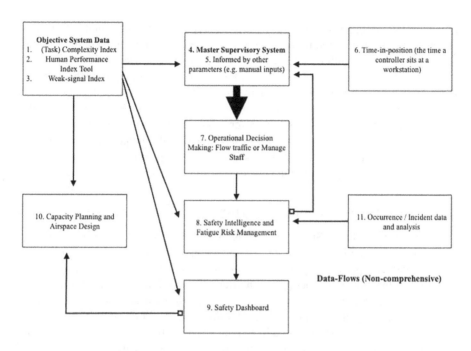

FIGURE 3.8.1 Overview of the Master Supervisor Decision-Support System (MSDSS).

and combines them with time-in-position data for each ATCO. It then calculates the best solution for how to use staff against the expected traffic and complexity of that traffic. For example, in busy summer months, Austria is regularly affected by large weather cells coming over the alps. This can dramatically change the complexity of the arriving and departing traffic in Vienna.

AI will be increasingly used to learn from previous data and solutions to provide even better recommendations. But there are many further benefits from this real-time operational support. Primarily they are safety intelligence, system performance post-operational knowledge, the ability to conduct safety incident investigations in a much more informed way, and the ability to design future capacity and airspace requirements.

The system components have mostly been assembled and each element is already proving to be of high value. Traffic complexity with enough sub-parameters and calculations is another excellent way of controlling the traffic and staff needed to manage it as it can easily be mined for historical learning and compared to weather and environmental impacts.

3.8.2 Evolving to an AI Driven System with Humans in the Loop

The approach outlined above is just the first step towards integrating various parameters to provide supervisory decision-making support. Eventually, the MSDSS could take over as the primary means of deciding how to optimize the airspace usage and the staff being used to control it. So long as the goal of the AI is to maintain humans-ITL as argued in Chapter 1, this would enable the individual ATCO who still hold licenses to maintain their situational awareness and complete their duties. If the MSDSS can maintain sustainable human performance against traffic and environmental demands then this ends up becoming a key part of the air traffic management concept of operation. The key here will be that the AI system does not let the operators get into situations that they cannot recognize or cannot cope with. Similar concepts can apply in the cockpits of aircraft and bridges of ships, but rather than a supervisor in an operations room, it would be the flight management system that controls how much the pilots have to do based on their human performance, fatigue levels, workload, and other human performance indicators.

3.8.3 CONCLUSIONS AND BUILDING TRUST

As of publication, the MSDSS concept is evolving as each component is coming online and being added. This is the case with implementing most AI solutions in aviation and transportation – that it will be done piece by piece and not in one go. But the benefits are already being shown after a limited time. Although the system is designed to keep humans in the loop and provide all the data necessary to understand what is going on by all humans, the trustworthiness of all of this data is still limited as operators adjust slowly to the adoption of these metrics. But as traffic demand continues to rise each year and operations around the world struggle to train enough ATCO fast enough, solutions like this might increasingly be required just to balance human performance against system demands.

4 Artificial Intelligence in Airport – Ports – Train Stations Operations

4.1 APPLICATIONS OF AI IN AIRPORTS – PORTS – TRAIN STATIONS MANAGEMENT, SECURITY, AND SURVEILLANCE SYSTEMS

Dimitrios Ziakkas

4.1.1 INTRODUCTION

The integration of Artificial Intelligence (AI) in transportation hubs such as airports, ports, and train stations has revolutionized how these critical infrastructures operate, manage crowds, ensure security, and provide services. AI's ability to look at huge amounts of data in real time and automate complicated tasks has huge benefits for improving security, business efficiency, and the overall travel experience (Ziakkas et al., 2023).

4.1.2 AIRPORTS

Automated Check-ins and Boarding Processes: AI-powered systems have streamlined the check-in and boarding process, reduced waiting times, and improved passenger flow through facial recognition and biometric verification.

Luggage Handling and Tracking: Advanced AI algorithms are used for efficient luggage sorting, routing, and tracking, significantly minimizing lost luggage incidents.

Security and Threat Detection Systems: AI enhances airport security through sophisticated surveillance cameras and anomaly detection algorithms that identify potential threats or unattended items in real-time.

Customer Service Enhancements: AI-powered chatbots and virtual assistants offer immediate information, flight notifications, and help travelers with questions, enhancing the customer care experience.

4.1.3 PORTS

Cargo Tracking and Management: AI systems enable real-time tracking and management of cargo, optimize logistics, and reduce bottlenecks in cargo movement.

Automated Crane Operations: AI-powered cranes improve loading and unloading efficiency, reducing manual labor and operational costs.

Security Patrols and Access Control: Drones and AI-based surveillance systems monitor port activities, enhancing security and access control measures.

Environmental Monitoring: AI applications in environmental monitoring help in tracking pollution levels and ensuring compliance with environmental regulations.

4.1.4 Train Stations

Ticketing and Access Control: AI facilitates automated ticketing systems and access control, improving operational efficiency and reducing queues.

Crowd Management: AI-powered surveillance systems monitor crowd densities and flow.

Security and Surveillance Operations: Enhanced security protocols through AI-driven surveillance, including facial recognition and behavior analysis, to ensure passenger safety.

Predictive Maintenance of Infrastructure: AI algorithms predict maintenance needs.

4.1.5 The Shipping Companies Case Study

Michail Diakomihalis

A key feature of shipping companies is their obligation to comply with the principles and rules governing the maritime transport market. Shipping companies operate under compliance regulations that apply to their day-to-day transactions. Therefore, what is required to achieve the highest compliance with the ISM code (International Maritime Organization (IMO), 2018) is the establishment of a daily organizational culture with practices for all employees and the company as an entity, to improve safety and promote the purpose of the business in the best way.

Commercial shipping operates according to documented guidelines, procedures, and regulations of international organizations. This framework is particularly positive for the application of AI since there is a large amount of data determined by the shipping market authorities, which is consistent with the role and work of AI. The AI system will be able to extract any information needed for any coded operation of the shipping company and the ship, wherever they are registered. For this reason, all information concerning the ship must be digitized and made available to AI, in such a way that for every question that can be asked, it is possible to search for and display the answer immediately.

The advantages of AI in the Shipping industry and the Improvements they bring may assist and improve the specific duties and responsibilities of the Operation Manager, which concern:

- The navigation and safety of operations on board are performed in accordance with the Company's regulations and the SMS manual.
- Compliance with the requirements of IMO, ICS, USCG, and other Maritime regulations.

AI can be applied in various sectors of the transportation industry to enhance outcomes, such as: (Poongavanam et al., 2023):

1. Fuel consumption, which at the same time contributes to the reduction of the environmental impact of the operation of ships.
2. Improvement of shipping network operations by adopting new technologies of AI with real business impact, such as auto-switching and auto-scaling hybrid cloud infrastructure in shipping businesses.
3. Empowering sustainable communities through smart technology in maritime and energy markets to create a digitally connected ecosystem with secure, intelligent, and cloud-based applications throughout the supply chain.

Specific effects that improve the shipping business include:

1. The use of advanced analytics.
2. The automation of operations.
3. Other applications of AI regarding security and ongoing assurance include cyber threats and malware that can be detected and countered with the support of machine learning.

The degree and evolution of the drag caused by pollutants on the in-water ship can be calculated by considering historical data from previous voyages of the ship.

REFERENCES

Poongavanam, S., Rajesh, D., Viswanathan, K., & Banu R. (2023). Role and Challenges of Artificial Intelligence in the Maritime Industry. *Journal of Survey in Fisheries Sciences*, 10(3S), 6313–6317

Ziakkas, D., Vink, L.-S., Pechlivanis, K., & Flores, A. D. (2023). *Implementation Guide for Artificial Intelligence in Aviation: A Human-Centric Guide for Practitioners and Organizations*. https://www.amazon.com/IMPLEMENTATION-GUIDE-ARTIFICIAL-INTELLIGENCE-AVIATION/dp/B0CL7Y4HPJ

4.2 THE USE OF AI IN BAGGAGE HANDLING, PASSENGER SCREENING, AND FACIAL RECOGNITION WITHIN AIRPORT, PORT, AND TRAIN STATION OPERATIONS

Anastasios Plioutsias

Combining AI in transportation hubs such as airports, ports, and train stations has dramatically improved operational efficiency and security. AI technologies are being increasingly adopted for baggage handling, passenger screening, and facial recognition, providing innovative solutions to traditional challenges (Oliveira et al., 2018). Automated sorting systems powered by AI algorithms quickly categorize and route baggage based on flight details, significantly reducing manual sorting errors. When combined with Radio-Frequency Identification (RFID) tags, AI offers real-time luggage tracking, providing a seamless flow from check-in to the aircraft (Wan et al., 2022). Predictive analytics help in forecasting baggage volumes, allowing for better resource allocation (Al-Ali et al., 2007). Adopting these technologies offers several benefits, including increased baggage handling accuracy, reduced loss, and misdirection rates, and enhanced operational efficiency and passenger satisfaction for airlines and airports (Heacock, 2005).

4.2.1 AI IN PASSENGER SCREENING

Advanced body scanners, powered by AI technology, can analyze images in real time to detect prohibited items without requiring manual image evaluation. This not only enhances privacy but also reduces wait times for passengers. New technologies like biometric verification systems like fingerprints and facial recognition have revolutionized passenger screening. Threat detection accuracy has been significantly improved by using machine learning algorithms to identify anomalies in passenger behavior or scanned images. Behavior detection algorithms powered by AI assess surveillance footage to identify any suspicious behaviors, enabling security personnel to respond proactively to any potential threats (Cardamo et al., 2010). These advancements guarantee a higher level of security and streamline the screening process, reducing wait times and drastically reducing the chances of human error. For example, Schiphol Airport in Amsterdam utilizes AI-driven scanners for passenger screening, which has significantly improved throughput and security measures.

4.2.2 AI IN FACIAL RECOGNITION

Facial recognition systems used at transportation hubs utilize AI to match individuals' faces against databases for identity verification and security checks. This technology enhances security by preventing identity fraud and expediting passenger flow through automated verification processes. It allows for speedy and accurate identification and verification of individuals, making boarding processes and access control smoother. Despite its benefits, facial recognition raises privacy and accuracy concerns, which necessitate strict regulatory compliance and ethical considerations. Hence, it is essential to implement this technology while keeping the ethical and legal aspects in mind. Facial recognition technology helps strengthen security and enables personalized experiences for passengers. It can facilitate tailored advertisements and automated check-ins, which redefine the concept of travel convenience. Atlanta's Hartsfield-Jackson Airport is an excellent example of the successful utilization of facial recognition technology for international travelers. It has significantly sped up the customs process while maintaining high security levels.

4.2.3 INTEGRATION CHALLENGES AND SOLUTIONS

Technical obstacles such as data integration, system compatibility, and ethical concerns like privacy and bias hinder the integration of AI technologies. A balanced approach prioritizing transparency and stakeholder engagement is required for successful implementation. Emerging technologies, such as quantum computing and advanced neural networks, promise to revolutionize AI's role in transportation further. The continuous improvement of AI algorithms and the integration of IoT

AI in Airport – Ports – Train Stations Operations 73

devices offer exciting prospects for enhancing security, efficiency, and passenger experiences.

4.2.3.1 Case Studies

The following case studies demonstrate how AI has been implemented in baggage handling systems with success. However, they also reveal some common challenges that come with it. These include the necessity of significant investments in technology and infrastructure, training staff to operate and maintain new systems, and ensuring the privacy and security of passenger data.

Case Study 1: Amsterdam Schiphol Airport

Schiphol Airport implemented an automated baggage handling system powered by AI. The system uses machine learning algorithms to optimize the routing and sorting of luggage. Real-time tracking uses RFID tags, ensuring accurate and efficient baggage processing for incoming and outgoing flights. The introduction of this system resulted in a significant reduction in baggage mishandling incidents, thereby improving passenger satisfaction and operational efficiency. Furthermore, the system's predictive maintenance capabilities helped minimize downtime, enhancing its effectiveness.

Case Study 2: Singapore Changi Airport

Singapore Changi Airport is recognized for its innovative approach to airport management. It deployed an AI-based baggage handling system integrating advanced image recognition technology. This technology identifies and sorts baggage based on size, type, and flight details. Additionally, the airport has complemented this system with autonomous robots that assist in transporting luggage to various parts of the airport. The cutting-edge system has expedited the baggage handling process and reduced the physical strain on human workers. The automation and precision provided by AI significantly improved the accuracy of baggage sorting. The result of this improvement is faster turnaround times for flights.

Case Study 3: Dubai International Airport

Dubai International Airport is one of the busiest hubs globally. It has implemented an AI-driven system to enhance its baggage handling operations. The system features sophisticated algorithms capable of predicting peak luggage volumes. This allows for proactive allocation of resources and optimization of baggage flow. The deployment of this AI system resulted in smoother operations, even during peak times, and a notable decrease in baggage handling errors. The airport also reported improved security measures, as the system includes features for automatic detection of suspicious or hazardous materials within luggage.

The case studies of Amsterdam Schiphol, Singapore Changi, and Dubai International airports show the potential of AI to enhance baggage handling processes. Using AI and machine learning, airports can boost operational efficiency, improve passenger experience, and ensure higher security.

REFERENCES

Al-Ali, A., Sajwani, F., Al-Muhairi, A., & Shahenn, E. (2007). Assessing the Feasibility of Using RFID Technology in Airports. In 2007 *1st Annual RFID Eurasia* (pp. 1–5). IEEE Xplore. https://doi.org/10.1109/RFIDEURASIA.2007.4368091.

Candamo, J., Shreve, M., Goldgof, D., Sapper, D., & Kasturi, R. (2010). Understanding Transit Scenes: A Survey on Human Behavior-Recognition Algorithms. *IEEE Transactions on Intelligent Transportation Systems*, 11, 206–224. Stamford: Key Publishing. https://doi.org/10.1109/TITS.2009.2030963.

Heacock, P. (2005). Improving Baggage Operations. *Airports International* (Vol. 38).

Oliveira, J., Nascimento, M., Júnior, J., & Freitas, C. (2018). RFID System Applicability Model for Traceability of Luggage at Airports. *International Journal of Advanced Engineering Research and Science*, 5, 264231. https://doi.org/10.22161/ijaers.5.8.16.

Wan, G., Li, W., Xu, Y., & Tong, M. (2022). A Novel Design of RFID System for Detecting and Sorting Baggage in Airports. In *2022 IEEE International Symposium on Antennas and Propagation and USNC-URSI Radio Science Meeting (AP-S/URSI)* (pp. 131–132). IEEE. https://doi.org/10.1109/AP-S/USNC-URSI47032.2022.9886431.

4.3 THE ROLE OF AI IN MULTIPLE REMOTE TOWERS AND GROUND OPERATIONS

Stefano Conte

4.3.1 AI SOLUTIONS FOR MULTIPLE REMOTE TOWER OPERATIONS

Remote tower operations involve providing aerodrome air traffic services remotely, replacing traditional control tower 'Out-the-Window view' (OTW) with high-resolution video panoramas from cameras on-site. This eliminates the need for physical towers, enabling Remote Tower Centres (RTC) to control multiple airports from one location, with each Air Traffic Controller (ATCO) managing one aerodrome at a time in a revolutionary concept known as Multiple Remote Tower Operations (MRTO). While remote towers are now a globally proven reality, MRTO is still in its incipient stage. Recently, extensive testing has been conducted in simulators and live exercises, producing contrasting results. This concept reduces personnel costs by flexibly allocating less ATCO workforce to control several airports, but it raises safety concerns because 'a single ATCO can perform the tasks originally designed to be executed by up to four ATCOs' (Li et al., 2018, p. 137).

These concerns prompted the International Federation of Air Traffic Controllers' Associations (IFATCA) to adopt the policy that 'ATCOs shall not be expected to provide a remote tower service for more than one aerodrome simultaneously' (IFATCA, 2018, p. 92). Therefore, simply developing technological enablers to control two or more airports is insufficient for implementing MRTO. It is crucial to establish a technically integrated, human-centered MRTO system to achieve successful operation. (Li et al., 2018). This is where the AI can make a difference. EASA proposes a three-level classification for AI deployment in aviation: Level 1 assists with human augmentation and cognitive assistance; Level 2 involves human-machine teaming with sublevels for cooperation and collaboration; and Level 3 envisions complete autonomy with user-overridable or non-overridable decisions (EASA, 2023). Human-AI Teaming, like team resource management, stresses shared understanding, trust, and effective interaction between human and synthetic members and could aid ATCOs in making quicker, informed decisions, boosting safety, and cutting delays (EASA, 2023). Research is also studying how AI could exploit its capability of instantly analyzing large amounts of data to predict events, something the human ATCO cannot do. This is the case of the SafeOPS project, which uses available ADS-B and weather data to train an AI that can predict the likelihood of an arriving aircraft performing a go-around.

4.3.2 AI Solutions for MRTO Challenges

Tools like SafeOPS can enhance ATCOs' situational awareness in MRTO, predicting potential overloading situations to help plan tactical solutions in advance and mitigate task load peaks. However, simultaneous MRTO poses unique human factor challenges, which AI can address. The most prominent one is certainly continuous attention allocation between concurrent tasks in different airports (Oehme et al., 2013), as the MRTO presents a substantial bottleneck in the allocation of the ATCO's visual attention while monitoring the traffic at multiple aerodromes (Papenfuß & Friedrich, 2016). The multiple resource theory states that when the perception channel (either visual or auditory) receives simultaneous competing stimuli, a sequential scanning between them is necessary; otherwise, interference and masking will occur. Research has shown that visual attention is controlled by a combination of processes: top-down which is expectancy-driven, and bottom-up, when attention is captured by stimuli (Wickens et al., 2022). Visual attention is closely tied to the physiological constraints of the eye. In MRTO, ATCOs must sequentially allocate visual attention to relevant information sources from multiple airports, risking the oversight of critical stimuli that would typically elicit bottom-up visual attention. It does not need to sequentially monitor the visual sources since it can manage unlimited concurring stimuli if the processing speed and memory capacity are sufficient. AI does not have focal vision or visual attention that must be rationed between sources. AI does not have limits to its perception; therefore, it has the potential to have a situational awareness that is always complete as long as the AI learns properly how to interpret its sources (e.g., OTW, ADS-B, weather data) and project the state of the system into the future.

In this context, a Level 2 AI would have the potential to be an all-seeing eye, supporting the ATCOs and summoning their attention in case the human visual attention is needed. For example, this kind of AI could autonomously monitor runways at different airports and draw attention in the event of a runway incursion at an airport that is not being actively watched by the ATCO. Experiments by Papenfuß and Friedrich (2016) on MRTO visual attention showed reduced monitoring of takeoffs and landings when managing multiple airports. Despite regulations requiring runway monitoring, this divided attention increases the risk of oversight. A level 2 AI could guide visual attention by signaling the relevant airport for ongoing operations and redirecting attention, and as the visual channel is highly exploited in MRTO, other sensory cues such as acoustic signals or haptic feedback could be used by AI to advise the ATCO. Another factor that increases the strain during MRTO is the repetitive mental switching between traffic situations at different aerodromes (Oehme et al., 2013). ATCOs develop specific and hierarchical task management patterns, but controlling multiple aerodromes simultaneously incurs a task-switching cost when shifting attention between airports. Factors such as uncertainty about the next task increase this cost. Clear stimuli specifying the task to perform can reduce cognitive load and reaction time (Wickens et al., 2022). Each airport has unique runways, taxiways, and procedures; thus, switching responsibilities needs mental reconfiguration. Interruptions mid-task increase the switching cost of this reconfiguration. If interrupted before finishing subgoals, tasks may fail. However, interruptions during expected breakpoints are less disruptive, suggesting that delaying interruptions could mitigate task-switching costs (Wickens et al., 2022).

A collaborative level 2 AI in MRTO could therefore reduce human error and ATCO's cognitive load by learning how to manage interruptions. Non-urgent events can be withheld by AI until the ATCO is between tasks, while urgent stimuli should be promptly presented with clear task cues to minimize uncertainty and switching costs. This concept could advance with a level 3 AI system autonomously deciding what is important enough to present to the ATCO. This kind of autonomous AI could be particularly useful when an abnormal or emergency occurs in one of the airports, as in such events the increased time pressure and increased workload would have a negative impact on ATCO's performance, even without external interrupting tasks (Kearney et al., 2018). Another challenge in MRTO for ATCOs is accurately allocating incoming information to the correct airport, a task critical for operational safety (Oehme et al., 2013). Many studies have investigated ways to avoid ATCO confusion between monitored airports. In 2018, the Shannon/Cork airports MRTO experiment used color-coded OTW displays to distinguish airports. However, the risk of "aerodrome confusion" persists, particularly during task execution as MRTO would require controlling airports with different layouts, dissimilar geographical features, and distinctive local procedures in a single shift. Those differences can lead to a fragmented situational awareness, which can potentially create misunderstandings, mix-ups, and other working errors. The similarity between ongoing and interrupting tasks can confuse, disrupting working memory with retroactive interference. After the interruption, lasting information may persist, causing proactive interference (Wickens et al., 2022). Such interferences could lead to an increase in rule-based mistakes, which occur when an action matches the planned objective but does not achieve it due to an improper application of a rule. In the MRTO case, the ATCO has greater potential to apply the correct procedure for a specific airport to the wrong one. To mitigate this, AI with automatic voice recognition could notify the

ATCO in real time if a message is sent to the wrong airport or contains instructions incompatible with that airport.

The use of AI in MRTO could also be extended to RTC supervisors. Flight plans and real-time surveillance positions could be easily analyzed by AI and level 3 automation could autonomously decide which airport will be assigned to each working position and manage rotations within shifts to optimize personnel utilization and avoid under- or overuse. AI could also learn to evaluate other factors in ATCOs/airport assignments that increase workload, such as severe weather (especially low visibility and strong winds), closure of parts of the maneuvering area, and proportion of instrumental and visual flights, like an experienced human supervisor.

4.3.3 Conclusion

The MRTO revolution opens up new ATM possibilities but also presents new obstacles. Concerns about those challenges led to IFATCA's MRTO veto. AI can support ATCOs with task execution, context awareness, and error recognition during MRTO. AI could help managers and ATCOs allocate airports and controllers in a flexible and analytic fashion after analyzing operational factors.

REFERENCES

EASA. (2023). AI ROADMAP 2.0 Human-Centric Approach to AI in Aviation. European Union Aviation Safety Agency. https://www.easa.europa.eu/en/document-library/general-publications/easa-artificial-intelligence-roadmap-20

IFATCA. (2018). *Technical and Professional Manual.* https://issuu.com/ifatca/docs/2018_01

Kearney, P., Li, W., & Braithwaite, G. (2018). Human performance assessment of multiple remote tower operations simultaneous Take-Off and landing at two airports. In *Lecture Notes in Computer Science.* https://doi.org/10.1007/978-3-319-91122-9\{_}47.

Li, W.-C., Kearney, P., Braithwaite, G., & Lin, J. J. H. (2018). How much Is too much on Monitoring Tasks? Visual Scan Patterns of Single Air Traffic Controller Performing Multiple Remote Tower Operations. *International Journal of Industrial Ergonomics*, 67, 135–144. https://doi.org/10.1016/j.ergon.2018.05.005.

Oehme, A., Leitner, R., & Wittbrodt, N. (2013). Challenges of Multiple Airport Control. *Aviation Psychology and Applied Human Factors.* https://doi.org/10.1027/2192-0923/a000034.

Papenfuß, A., & Friedrich, M. (2016). Head Up Only – A Design Concept to Enable Multiple Remote Tower Operations. In *35th Digital Avionics Systems Conference*, Sacramento, CA. https://doi.org/10.1109/DASC.2016.7777948.

Wickens, C. D., Gutzwiller, R. S., & McCarley, J. S. (2022). *Applied Attention Theory* (2nd ed.). CRC Press, Boca Raton. https://www.taylorfrancis.com/books/9781003081579.

4.4 ARTIFICIAL INTELLIGENCE IN SHIPPING OPERATIONS
Michail Diakomihalis

A key feature of shipping companies is their obligation to comply with the principles and rules governing the maritime transport market. Specifically, compliance is required for the characteristics of each ship relating to the flag of the country they fly and where the ship is registered, the legislation of the states that the ship approaches and carries out transports, the requirements of the charterers, and the obligations of the company according to the legislation of the country where the ship is registered. Shipping companies operate under compliance regulations that apply to their day-to-day transactions. Therefore, what is required for achieving the highest compliance with the ISM code (International Maritime Organization (IMO), 2018) is the establishment of a daily organizational culture with practices for all employees and for the company as an entity, with the aim of improving safety and promoting the purpose of the business in the best way.

Commercial shipping operates according to documented guidelines and procedures, and regulations of international organizations. The AI system will possess the capability to retrieve any required information on the coded operations of the shipping firm and the ship, regardless of their registration location. Hence, it is crucial to digitize and provide AI with comprehensive access to all ship-related data, enabling instant retrieval and display of answers to any possible inquiry.

The Operations Manager ensures that the management department's concepts and objectives are implemented and adhered to, and he or she approves and executes purchases up to the General Manager's budget. He reports to the General Manager, as do all of his colleagues and subordinates in his department. AI in the shipping industry provides several benefits and advancements that can help optimize the Operations Manager's activities and responsibilities. These include:

- The navigation and safety of operations on board are performed in accordance with the Company's regulations and the SMS manual.
- Oversee purchases of deck and engine room supplies (except spare parts and engineering supplies).
- Complies with the requirements of IMO, ICS, USCG, Maritime regulations, procedures and requirements of the Company's SMS.
- Collaborates with the Company's Charter and Insurance departments.
- Processes vessel voyage instructions and oversees their progress in accordance with the requirements of charter agreements.
- Appoints Vessel Agents to fulfill C/P requirements.
- Advises Masters on C/P matters.
- Monitors the execution of the voyages and the operation of the ships on the voyage and in the port.
- Compiles and processes voyage documents, manages fuel economy and regulates ships' fuel and water supply.
- Collaborates with other departments on the ship's performance, maintenance, and operational status requirements.
- The Director of Supplies acts jointly and with the cooperation of the other Departments to achieve the principles and objectives of the Company, and signs and binds the Company in matters of his Department's competence. The main responsibilities are:
- Keeping records for the consumption of supplies on the ships, cooperation with the ships, and the local agents for the orders of food, water, etc.
- Arrangement and purchase of engine and deck supplies in cooperation with the responsible departments and processing of the documents.

- Diligence and control of the hotel supplies on the ship.
- Refers to SMS DPA for cost control and consumption issues.
- Evaluation of Suppliers.

Every new technology affects the operation of businesses and the market. What is expected, therefore, is the effect of AI on shipping companies and shipping in general. Among the applications that affect the shipping industry is the reduction of fuel consumption, resulting in a reduction of travel costs and ultimately the improving of the operation of the business. The vessel should strive to reach its optimal Energy Efficiency of Operation, which is defined as a metric that demonstrates the ratio of fuel consumption by the main engine per unit of transport effort (Lu et al., 2015).

AI can be used in different areas of the shipping business for better results, as descibed by Poongavanam et al. (2023):

- The reduction of fuel consumption has been implemented in large shipping companies as a result of the AI application, which at the same time contributes to the reduction of the environmental impact of the operation of ships.
- Improvement of shipping network operations by adopting new technologies of AI with real business impact, such as auto-switching and auto-scaling hybrid cloud infrastructure in shipping businesses.
- Empowering sustainable societies with smart technologies for marine and energy markets results in the creation of an ecosystem that is digitally connected throughout the supply chain through applications that are secure, intelligent, and cloud-based.

AI is a technology that utilizes historical company data to analyze and recommend optimal decisions for the future of the firm. The support provided by AI applications in shipping is crucial for decision-making and operational matters, such as automation, safety, selecting the best travel routes, and improving ship efficiency.

4.4.1 FUTURE OPTIONS

- Utilizing advanced analytics enables the extraction of useful business insights from many data sources, thereby ensuring that the decisions taken by company executives are grounded in validated methodologies.
- Shipping businesses prioritize the automation of operations, ship equipment, and communications. Machine learning is accomplished by analyzing historical occurrences, such as variations in weather, periods of high activity in the maritime transportation industry, periods of economic downturn in load finding, etc.
- Other applications of AI regarding security and ongoing assurance include cyber threats and malware that can be detected and countered with the support of machine learning. With this feature, unwanted accidents in the transfer, download and storage of information and data can be reduced.
- The optimization of the ship's route for a voyage can be done through the processing of data that has been integrated and predictive models created with the application of AI.

REFERENCES

International Safety Management (ISM) Code. (1993). International Management Code for the Safe Operation of Ships and for Pollution Prevention, Assembly – 18th Resolution A.741(18). https://www.imo.org/en/ourwork/humanelement/pages/ISMCode.aspxbot

Lu, R., Turan, O., Boulougouris, E., Banks, C., & Incecik, A. (2015). A Semi-Empirical Ship Operational Performance Prediction Model for Voyage Optimization Towards Energy Efficient Shipping. *Ocean Engineering*, 110, 18–28.

Poongavanam, S., Rajesh, D., Viswanathan, K., & Banu R. (2023). Role and Challenges of Artificial Intelligence in the Maritime Industry. *Journal of Survey in Fisheries Sciences*, 10(3S), 6313–6317.

4.5 TRAIN OPERATIONS

Beatrix Walzl

AI technologies have emerged as powerful tools for transforming various aspects of train station operations, offering numerous applications aimed at enhancing efficiency, safety, and passenger experience. Tang et al. (2022) conducted an extensive literature review of AI in railway systems and focused on five sub-domains: Maintenance and Inspection, Traffic Planning and Management, Safety and Security, Autonomous Driving and Control, and Passenger Mobility. They conclude that the main scientific research has looked at Maintenance and Inspection (58%), followed by Traffic Planning and Management (24%). All other sections have received limited attention (Tang et al., 2022). This chapter will focus on the role of AI in train station operations with regard to Traffic Planning and Management, as it provides a decent amount of research as of 2022 after Maintenance and Inspection.

4.5.1 Applications of AI in Train Operations

4.5.1.1 Re-Scheduling and Re-Allocation of Resources

Only when long-distance trains become a reliable transport system will trains be a real competitor to short-haul flights. AI has the potential to decrease the impact of disruptions and increase the passenger experience in regard to the need for punctual and reliable railway transportation. AI algorithms can analyze historical data, current traffic conditions, and passenger demand to optimize train schedules, minimize delays, and re-allocate resources efficiently. For example, AI-powered systems can dynamically adjust train timetables in response to unexpected disruptions or changes in demand, ensuring smoother operations and better utilization of railway infrastructure (Deutsche Bahn, 2023a). Tang et al. (2022) brought together multiple studies showing solutions for rescheduling and the handling of disruptions, which they distinguish between train-oriented approaches such as minimizing delays (e.g., Wang et al., 2019) and what they call passenger satisfaction-oriented approaches to maximize the quality of the passenger experience (e.g. Obara et al., 2018). For a train-oriented approach to decreasing delays, Wang et al. (2019) propose a particle swarm optimization algorithm. One of their constraints is the consideration of not only delays but serious disruptions such as complete blockages. Obara et al. (2018) concluded with a similar issue: their model is not capable at the moment of handling large-scale delays.

4.5.2 Challenges in Implementing AI in Train Operations

Implementing AI in train operations presents various challenges that must be addressed to ensure successful integration and adoption. AI is still in the early stages of being implemented in the railway industry. One of the main challenges is that the lifecycles of trains and train stations can be quite extensive, e.g., 20–40 years or even longer (Sasidharan et al., 2020). Therefore, a railway operation must handle technologies and engineering that might be well over 20 years old. This can cause issues with gathering data, data integration and the availability of big data, especially when talking about long-distance trains that might even go through different countries, such as in the European Union. Gathering reliable data would entail receiving valid and real-life data from a range of providers that work with different systems, in different countries, languages, legal boundaries, and approaches.

The European Union Agency for Railways was established in 2006, and they have since worked on implementing common technical standards, signaling standards, and safety measures. This is a lengthy process, that will be ongoing for many years or even decades. This means the railway industry is forced to work within the constraints of a slow transition to common standards that would deliver a reliable set of big data across borders for AI-supported tools to increase efficiency and quality. Therefore, the use of AI within railway systems is currently constrained to local and independent networks (e.g., Deutsche Bahn, 2023b). Considering an overall goal to maximize the quality of the end-user (the passenger) experience, which means high reliability, a decrease in delays, an optimized flow between connections, and the availability of alternate routes or options when disruptions occur, there is a general lack of broader human-centric design approaches and standards for the entirety of a railway system. Other global industries, such as the aviation and maritime industries, have standardized their signaling, processes, etc. over decades, which provides an easier basis for big data analysis and holistic system views. An extensive amount of research will be needed to smoothly use AI tools for railway systems and make full use of the advantages of AI.

4.5.3 Outlook and Conclusions

AI technologies hold immense promise for transforming train station operations, particularly in the domains of traffic planning and management, as well as passenger mobility. By leveraging AI algorithms, train operators can optimize train schedules, minimize delays, and enhance passenger flow within train stations, ultimately improving the overall efficiency and reliability of railway transportation systems, as well as the passenger experience. However, challenges such as data integration, privacy concerns, and resistance to change need to be addressed to ensure the successful implementation and adoption of AI in train station operations. Looking ahead, continued research and innovation in AI-driven solutions within railway systems are essential to realizing the full potential of AI in shaping the future of railway transportation.

The integration of AI technologies holds immense potential for revolutionizing train operations, particularly in areas such as traffic planning and management (Xie et al., 2020). By deploying AI algorithms, railway operators have the opportunity to optimize schedules, minimize delays, and enhance passenger flow, thus significantly improving the overall efficiency and reliability of railway transportation systems. However, the successful implementation of AI in train operations is accompanied by various challenges, such as data integration across disparate systems.

AI in train operations has a bright future despite these obstacles. Railway operators can overcome these challenges and realize AI's transformational potential by investing in research and innovation, promoting stakeholder engagement, and prioritizing human-centric design. AI integration might transform the railway sector, making it safer, more efficient, and more passenger-focused.

REFERENCES

Deutsche Bahn. (2023a). Erläuterung Pünktlichkeitswerte für den Januar 2024. DB Konzern. https://www.deutschebahn.com/de/konzern/konzernprofil/zahlen_fakten/puenktlichkeitswerte-6878476.

Deutsche Bahn. (2023b, June 18). Deutsche Bahn weitet Einsatz von KI für pünktlichere Züge aus. DB Konzern. https://www.deutschebahn.com/de/presse/pressestart_zentrales_uebersicht/Deutsche-Bahn-weitet-Einsatz-von-KI-fuer-puenktlichere-Zuege-aus-10771280.

Obara, M., Kashiyama, T., & Sekimoto, Y. (2018, December). Deep Reinforcement Learning Approach for Train Rescheduling Utilizing Graph Theory. In *2018 IEEE International Conference on Big Data (Big Data)* (pp. 4525–4533). IEEE. Doi: 10.1109/BigData.2018.8622214

Sasidharan, M., Burrow, M. P. N., & Ghataora, G. S. (2020). A Whole Life Cycle Approach Under Uncertainty for Economically Justifiable Ballasted Railway Track Maintenance. *Research in Transportation Economics*, 80, 100815.

Tang, R., De Donato, L., Besinović, N., Flammini, F., Goverde, R. M., Lin, Z., ... & Wang, Z. (2022). A Literature Review of Artificial Intelligence Applications in Railway Systems. *Transportation Research Part C: Emerging Technologies*, 140, 103679.

Xie, P., Li, T., Liu, J., Du, S., Yang, X., & Zhang, J. (2020). Urban Flow Prediction from Spatiotemporal Data Using Machine Learning: A Survey. *Information Fusion*, 59, 1–12

5 Artificial Intelligence in Customers' Experience

5.1 AI-BASED CHATBOTS AND VIRTUAL ASSISTANTS FOR PERSONALIZED CUSTOMER INTERACTIONS

Anastasios Plioutsias and Dimitrios Ziakkas

AI-based chatbots and virtual assistants are changing the way businesses communicate with customers. These tools simulate human conversation through text chats or voice commands, providing customers with a personalized and efficient service experience. They are essential as they offer a scalable and cost-effective solution to meet the increasing demand for instant and on-demand customer support. In the transportation industry, where customer satisfaction is crucial, airlines and airports adopt AI-based chatbots and virtual assistants. These technologies offer personalized, efficient, and scalable solutions to meet the diverse needs of travelers. These tools improve customer service interactions by utilizing advanced AI, making them more responsive and tailored to individual passenger preferences.

5.1.1 Evolution of Chatbots and Virtual Assistants

The development of chatbots and virtual assistants has come a long way. Initially, chatbots could only respond to specific commands or phrases with pre-programmed responses. However, with the advancement of AI technology, these systems can now understand and process natural language, engage in contextual conversations, and learn from interactions to improve their performance over time. A The AI-powered solutions are used across various touchpoints in the passenger journey, providing 24/7 support, personalized travel information, and seamless service interactions (Prajwal et al., 2019).

5.1.2 Technologies Behind AI Chatbots and Virtual Assistants

NLP enables chatbots to understand human language, identify intents, and respond contextually. Virtual assistants can answer client questions naturally and conversationally (Toader et al., 2019). Nowadays, chatbots can better answer questions using Machine Learning and Deep Learning since they can learn from data, find patterns, and make choices. Data-learning technology improves responses and personalizes interactions depending on passenger preferences and habits (Prajwal et al., 2019). Finally, virtual assistants interpret and act on speech commands using voice recognition technologies, making interactions easier (Shafeeg, 2023). This technology allows hands-free communication, making travel easier.

5.1.3 Applications in Customer Service

AI chatbots and virtual assistants provide round-the-clock support in various customer service scenarios. They can answer queries and resolve issues anytime, and their multilingual capabilities make it possible to serve customers in their native language (Dimitriadis, 2020). By analyzing customer data, personalized recommendations can be made to enhance the shopping experience and support. The AI-driven assistants can take care of automated check-ins and provide real-time flight updates, notifying passengers of gate changes, delays, and boarding times through personalized travel updates. Additionally, they can handle customer queries and offer a hassle-free way to manage bookings, including changes and cancellations. Finally, chatbots can offer immediate assistance in answering FAQs and resolving complex travel issues, reducing wait times and improving customer satisfaction.

5.1.4 Benefits for Businesses and the Aviation Industry

Using AI-based chatbots offers several advantages, such as streamlining operations, reducing the need for human agents, and lowering costs (Dimitriadis, 2020). Furthermore, personalized, real-time interaction enhances customer satisfaction and loyalty, improving the passenger experience. Automating routine inquiries also improves operational efficiency, enabling staff to focus on higher-value tasks and reducing operational costs. Instant, 24/7 interaction data provides valuable insights into customer preferences and behavior, enabling airlines to tailor services and offers to meet their needs better.

5.1.5 Challenges and Considerations

The privacy and security of customer data should be a top priority. It is essential to seamlessly connect AI systems with airline and airport databases for accurate, real-time information. Safeguarding personal and sensitive passenger data within AI interactions is a primary concern, requiring robust security measures.

5.1.6 Future Trends

Integrating AI with IoT devices will also enable a more connected and intuitive user experience. Finally, there will be a focus on ethical AI use and enhancing human-AI collaboration. Finally, leveraging AI for real-time advisories and support during travel disruptions can help manage crises.

REFERENCES

Dimitriadis, G. (2020). Evolution in Education: Chatbots. *Homo Virtualis*, 3, 47–54. https://doi.org/10.12681/homvir.23456.

Prajwal, S., Mamatha, G., Ravi, P., Manoj, D., & Joisa, S. (2019). Universal Semantic Web Assistant Based on Sequence to Sequence Model and Natural Language Understanding. In *2019 9th International Conference on Advances in Computing and Communication (ICACC)* (pp. 110–115). https://doi.org/10.1109/ICACC48162.2019.8986173.

Shafeeg, A., Shazhaev, I., Mihaylov, D., Tularov, A., & Shazhaev, I. (2023). Voice Assistant Integrated with Chat GPT. *Indonesian Journal of Computer Science*. htttps://doi.org/10.33022/ijcs.v12i1.3146.

Toader, D., Boca, G., Toader, R., Macelaru, M., Toader, C., Ighian, D., & Rădulescu, A. (2019). The Effect of Social Presence and Chatbot Errors on Trust. *Sustainability*. https://doi.org/10.3390/su12010256.

5.2 USE OF AI RECOMMENDATION SYSTEMS, PRICING OPTIMIZATION, AND REVENUE MANAGEMENT

Volodymyr Bilotkach

5.2.1 INTRODUCTION

Since the Great Recession, airline revenue structures have evolved significantly, largely due to product unbundling and the rise of 'ultra-low-cost carriers' (ULCCs). This has led to ancillary revenues becoming a crucial income source, with charges for extras like checked baggage, seat selection, and more. The integration of AI in revenue management (RM) and ancillary services, such as third-party product sales and frequent flier programs (FFP), presents a potential for enhancing profitability. The application of AI could streamline operations and improve customer loyalty through more personalized and efficient service offerings. Pricing of ancillary revenues is currently done in a rather crude and unimaginative way. Flat charges for checked baggage are the norm, with many carriers offering discounts for purchasing this service online rather than at the airport, and some airlines differentiate checked baggage charges by weight and/or length of haul. There is some differentiation by seat quality in the pricing of advance seat reservations.

The traditional RM approach in the airline industry involves the separation of the price-setting function from that of dynamic capacity allocation. The determination of various fare classes, including both setting airfares and determining various fences, is the purview of the airline's Pricing Department. The job of a RM department has been that of dynamic capacity allocation – evaluating which booking classes were to remain open or closed, depending on the discrepancies between expected and realized booking levels. Whether AI can lead to a better integration of the two functions remains an open question., In my opinion, the key constraint here remains the computing power required to solve the complex underlying optimization problem. Yet, this new technology is well poised to bring improvements within the current RM paradigm.

Other areas where AI could be used to enhance revenue opportunities for the airlines are sales of commission-based products offered by third parties (most notably, insurance and services complementary to air travel, such as hotels and rental cars) and FFPs. FFPs, used to create and maintain a pool of brand-loyal customers, also serve as an important ancillary revenue generator for airlines, especially in North America, through partnerships with financial institutions. Sorensen (2022) notes that the three largest airlines in the USA have generated over $12 billion in revenues through their FFPs. The key problem the airlines need to solve as far as FFPs are concerned is making sure customers redeem their points in a manner that minimizes cost to the airline, including the opportunity cost of seats frequent fliers buy with miles. I see quite a lot of potential for AI in this area.

5.2.2 Opportunities for Traditional RM

Airlines' RM systems work in the following general fashion: The Pricing Department is responsible for determining various fare categories for all the origin-destination markets the airline is operating in. Once the multitude of price categories has been set, the RM department is responsible for determining which of the price categories are to be made available for booking (and, in some cases, how many seats to allocate to different price categories) for each origin-destination market and/or flight on any given day. This process is known as dynamic yield management. Yield management operates under the simple general principle of comparing the expected number of bookings with the actual number and responding in case of discrepancies in either way (when you have more or fewer bookings than you expect on a given flight at a given time) by adjusting the allocation of seat capacity across fare levels and/or opening or closing price categories to achieve a better outcome.

The airlines' RM systems are set up to maximize either total revenue or passenger load factor. The latter is more common for the LCCs and ULCCs, as these carriers rely to a greater extent on revenues from add-on services. Optimization of the RM objective function is a complex process, especially for the airlines operating hub-and-spoke networks. The sheer complexity of this problem makes solving it a time-consuming exercise, even with the current computing capabilities. Solving the joint pricing-capacity-allocation problem is a task that most airlines cannot address, due to its complexity and multi-dimensionality. Thus, we would not expect AI to enable putting Pricing and RM departments under the same roof for most airlines.

Nevertheless, AI can be used to improve both pricing and RM processes. The key opportunities for the use of AI here are in improving the systems' predictive capabilities. This will allow the RM systems to flag cases of bookings being significantly below or above the expected levels before the gap between actual and expected bookings becomes significant. This can be accomplished by training AI models on past data to help them identify the warning signs typical of cases that eventually require revenue managers' intervention. Airlines can also use the data on outcomes of past interventions by the revenue managers to generate suggestions on the action's managers can take, complete with information about the likely outcomes of such actions. Monitoring and predicting competitors' actions is another task where AI can help.

What the general public sees as ever-changing airfares is actually a reflection of changes in the availability of different price categories. The prices associated with different categories themselves change relatively infrequently. By monitoring changes in offered fares, AI can detect and alert to apparent changes in competitors' price levels, thus providing valuable insights to the Pricing Department and higher-level managers.

5.2.3 Optimizing Add-On Pricing and Sales

Add-on service pricing is an area of RM that can be improved. Airlines that depend on add-on services for a large portion of their revenue keep base fares low to attract customers, then offer a variety of additional services at a fixed price, with some differentiation by length of haul or weight for checked baggage fees. Similar dynamic

Artificial Intelligence in Customers' Experience

seat capacity management methods can be used here. An airline may provide cheaper checked baggage fees for business flights and higher rates on leisure trips, with the lowest fees available when buying add-ons upon booking. This procedure can be improved by AI data summarization.

AI models can also advise pricing changes for seat selection, pre-booked meals, and other add-on services based on client data. AI can leverage RM model data to give customers booking advice. AI-generated messages delivered at the proper time may boost add-on sales. Bundles of add-on services are now routinely offered by the LCCs and ULCCs. These bundles typically include a multitude of services and offer discounts as compared to purchasing components of the bundle separately. AI may help airlines increase the number of customers opting for those bundles by nudging them through the messages.

An important requirement for the effectiveness of such messages is establishing credibility with the customers. Moreover, adding a note that the message was indeed generated by an AI engine external to the airline could help establish such credibility.

5.2.4 USING AI FOR SALES OF COMMISSION-BASED PRODUCTS AND ADVERTISING

While commission-based product sales (most notably, products complementary to air travel, such as hotel and rental car bookings, as well as various types of travel-related insurance) are not a big driver of airline revenues, AI does offer a chance to significantly improve this aspect of revenue generation. Moreover, AI can enhance the offerings presented to the customer. The customer can play either a semi-passive or an active role here. The options can be framed in a more interactive way than what one expects to see when booking through online travel agents. When a customer takes on an active role, he/she is taken to an AI-powered tool, where the passenger can describe his/her requirements for a hotel or rental car using a ChatGPT-type query. A dialogue between the customer and the AI tool would ensue, resulting hopefully in the customer making the bookings and the airline receiving commission.

Advertising is another small but not insignificant revenue source for the airline. With an increasing number of airlines offering in-flight entertainment (IFE) through streaming on passengers' devices, the opportunities for selling advertising via IFE are increasing. AI-powered solutions can summarize passengers' IFE interactions, giving airlines more complete and up-to-date information.

5.2.5 GETTING MORE OUT OF FFPS

Airlines' customer loyalty programs (known as frequent flier programs, or FFPs) have become an integral part of the industry. While some carriers (most notably, European LCCs) shy away from offering such programs, most airlines do not. Through partnerships with financial institutions, many airlines (especially in North America) have turned their FFPs into revenue generators.

FFP has to provide enough value for the customer to exhibit loyalty to the airline by occasionally paying more to fly with the preferred carrier as compared to what the competitors might charge without imposing too much of a cost on the airline. Points accrued by the loyal customers represent liability to the carrier in the form of free seats that could have been occupied by fare-paying passengers and the cost of providing various perks of the FFP membership and status (free checked bags and free or discounted add-on services, lounge access, etc.).

AI can be used in this area to further improve upon the developments that have already been made. The data from the airline's FFP can be used to predict demand for award tickets on specific flights and dates and to optimize the 'pricing' of those tickets. On the redemption side, AI-powered tools can be developed to suggest to the customers ways of spending their points. As with the sales of products complementary to air travel, the customer can take either a semi-passive or active role here. The tools can suggest award flight options and other ways to take advantage of the points the customer has accumulated, in an interactive fashion, similar to ChatGPT dialogue. It should not be difficult to incorporate such tools into the airlines' websites or apps.

To conclude, the use of AI-generated messages seems to be a promising avenue for utilizing this new technology to provide additional information to customers. The key problem to be overcome when attempting to use those messages at the time the customer is booking travel is establishing trust. The messages need to be perceived as providing relevant information that helps the passenger make a decision that maximizes the value for money from the booking rather than trying to convince him/her to spend more money.

REFERENCE

Sorensen, J. (2022). *The 2022 CarThrawler Yearbook on Ancillary Revenue.* IdeaWorks Company.com, Shorewood, WI, USA

5.3 THE ROLE OF AI IN CUSTOMER FEEDBACK ANALYSIS, SENTIMENT ANALYSIS, AND SOCIAL MEDIA MONITORING

Caterina Ciacatani

In the digital age, organizations are realizing the importance of customer feedback, sentiment analysis, and social media monitoring to better understand and address consumer opinions. AI has become a potent tool, providing a sophisticated ability to analyze extensive data and derive significant insights. The presented analysis delves into the role of AI in *customer feedback analysis, sentiment analysis, and social media monitoring,* exploring the benefits, challenges, and prospects of integrating AI into these critical business processes.

5.3.1 CUSTOMER FEEDBACK ANALYSIS

The efficient processing and organization of large volumes of customer feedback data represent a cornerstone in the utilization of AI within the domain of customer feedback analysis AI, particularly NLP algorithms, has transformed how businesses handle and get insights from unstructured textual data. When it comes to the role of AI in customer feedback analysis, several key aspects stand out:

Artificial Intelligence in Customers' Experience

- Data Collection and Aggregation
- Natural Language Processing (NLP)
- Sentiment Analysis: AI-driven sentiment analysis delves deep into customer emotions, offering businesses a nuanced understanding of how customers perceive their products or services (Martin, 2019).
- Categorization and Trend Identification
- Predictive Analytics
- Continuous Learning and Improvement
- Integration with Business Systems

AI-powered customer feedback analysis seamlessly integrates with CRM and business intelligence systems, providing a comprehensive understanding of customer interactions and influencing strategic decisions (Jurafsky & Martin, 2019). Moreover, the transformative role of AI in customer feedback analysis empowers businesses to refine their products, services, and strategies in alignment with evolving customer expectations. As technology progresses, the integration of AI is poised to become even more sophisticated, offering endless opportunities for businesses to leverage customer insights and enhance their operations.

5.3.2 Sentiment Analysis

A technique of NLP, sentiment analysis, alternatively referred to as opinion mining, executes sentiment or emotive tone extraction from a given text. AI significantly contributes to *sentiment analysis* by introducing sophisticated functionalities that facilitate the comprehension and interpretation of human emotions expressed in textual data. The integration of AI in sentiment analysis has significantly enhanced the ability of businesses to interpret and respond to customer feedback on a nuanced level. AI leverages *Natural Language Processing (NLP)* techniques (Automated Text Processing) to understand human language intricacies, including grammar and context, enabling it to analyze sentiments with a degree of accuracy that mirrors human interpretation. Techniques such as *tokenization and part-of-speech tagging* further refine this analysis, allowing for a detailed understanding of sentiment at the word and sentence level. Moreover, AI's capability extends to *multilingual sentiment analysis*, thanks to advanced models trained on diverse datasets (language agnosticism), facilitating *cross-lingual sentiment transfer,* and enabling businesses to gather insights from global customer bases without language barriers. Beyond mere positive or negative sentiment classification (contextual understanding /context-aware analysis), AI delves into fine-grained emotion detection, identifying a wide spectrum of emotions such as joy, anger, and surprise, and even assessing the intensity of these emotions (emotion detection). AI can identify sentiments not only at the document level but also at the entity or aspect level. This means that sentiments towards specific entities or aspects within a text, such as products or features, can be individually analyzed. This detailed analysis is crucial for businesses aiming to understand the depth of customer emotions towards products or services. AI's *adaptability* and continuous learning ensure that sentiment analysis tools stay relevant amidst evolving language patterns, making AI an indispensable asset for *real-time sentiment monitoring and predictive analytics*. This transformative role of AI not only offers businesses a sophisticated approach to gauging customer sentiment but also promises to become more integral and accurate as technology advances, ensuring that businesses remain closely aligned with their customers' emotions and opinions (customization and adaptability).

5.3.3 Social Media Monitoring

Individuals, businesses, and communities now discuss, share, and shape narratives on social media. AI has become a key player in social media monitoring due to the exponential expansion and complexity of social media data. AI technologies have transformed how organizations collect information, manage brand reputation, and communicate with their audience from laborious to efficient, scalable, and real-time. Voice-activated virtual assistants make interactions easier. This technology allows hands-free communication, making travel easier. Through leveraging advanced NLP algorithms, AI enables a deeper dive into sentiment analysis, moving past simple positive or negative classifications to identify nuanced emotions behind user-generated content (Jurafsky & Martin, 2019). This enhanced emotional intelligence allows businesses to refine their strategies and responses, tailoring them with unprecedented precision to align with public sentiment.

As AI continues to evolve, its application in social media analytics becomes increasingly integral, offering a multi-dimensional approach that encompasses *efficient data processing, scalability, real-time insights, contextual understanding, and continuous learning and adaptation*. The *challenges and ethical considerations* inherent in deploying AI technologies underscore the importance of navigating this landscape thoughtfully (Martin, 2019).

5.3.4 Conclusion

AI's role in customer feedback analysis, sentiment analysis, and social media monitoring is transformative, offering businesses unprecedented insights into customer perceptions (Davenport & Ronanki, 2018). As technology continues to evolve, companies that harness the power of AI in these areas will be better positioned to understand their customers, enhance brand loyalty, and stay ahead in the competitive landscape (Caliskan et al., 2017).

REFERENCES

Caliskan, A., Bryson, J. J., & Narayanan, A. (2017). Semantics Derived Automatically from Language Corpora Contain Human-like Biases. *Science*, 356(6334), 183–186. https://doi.org/10.1126/science.aal4230.

Davenport, T. H., & Ronanki, R. (2018, January 9). Artificial Intelligence for the Real World. *Harvard Business Review (HBR)*. https://www.scirp.org/(S(czeh2tfqw2orz553k1w0r45))/reference/referencespapers.aspx?referenceid=3166319

Jurafsky, D. & Martin, J. H. (2019). *Speech and Language Processing. Scientific Research Publishing*. https://www.scirp.org/(S(351jmbntvnsjt1aadkozje))/reference/referencespapers.aspx?referenceid=2976238

5.4 THE AIRLINERS' CASE STUDY
Anastasios Plioutsias and Dimitrios Ziakkas

The aviation industry uses AI to enhance the customer experience. The case study demonstrates how airlines implement AI to offer personalized and efficient customer service technologies. This helps cater to customer touchpoints by providing tailored recommendations and a deeper understanding of their preferences, resulting in a highly personalized travel experience. The study emphasizes the benefits of AI in enhancing customer satisfaction, streamlining operational efficiencies, and encouraging innovation. The aim of this case study is to provide a comprehensive overview of AI in aviation customer service, showcasing innovative solutions and practices that can revolutionize the customer experience.

5.4.1 Background

Airlines are incorporating AI technologies to improve customer service. AI simplifies the pre-flight process, tailors onboard experiences, and automates feedback collection. Implementing AI comes with technical challenges, such as integrating with existing systems and ensuring data privacy. Airlines have addressed these challenges with strategic planning and phased implementation.

5.4.2 Technical Integration with Legacy Systems

Integrating AI with legacy systems is complex and resource-intensive for many airlines. Airlines are adopting middleware solutions and gradually replacing outdated systems with AI-compatible platforms.

5.4.2.1 Ensuring Data Privacy and Security
AI in customer service presents privacy and security concerns. Data protection laws like the GDPR complicate things. Airlines protect client data using enhanced cybersecurity, data anonymization, end-to-end encryption, and secure storage.

5.4.2.2 Balancing Automation with Human Touch
AI improves efficiency and personalization but may reduce customer service. Balancing automation and human connection is key (Ziakkas et al., 2023). A hybrid strategy allows airlines to use AI for everyday activities and human operators for complicated issues and personal care. AI is effortlessly integrated with staff training. Furthermore, AI has raised customer expectations for personalized and efficient service, which is challenging to provide consistently. Airlines manage expectations by being transparent about AI capabilities and collecting feedback to improve services.

5.4.3 Outcomes

AI has greatly improved the aviation industry, benefiting customer satisfaction, operational efficiency, and employee productivity. Quantitative outcomes of AI include a 35% reduction in unexpected downtime, a 25% increase in passenger satisfaction, and significant cost savings for airlines. Moreover, qualitative outcomes of AI include enhanced customer loyalty and employee satisfaction, as well as the introduction of new services such as real-time luggage tracking and AI-powered IFE recommendations.

REFERENCE

Ziakkas, D., Vink, L.-S., Pechlivanis, K., & Flores, A. (2023). *Implementation Guide for Artificial Intelligence in Aviation: A Human-Centric Guide for Practitioners and Organizations*. HFHorizons. ISBN 9798863704784. https://www.amazon.com/IMPLEMENTATION-GUIDE-ARTIFICIAL-INTELLIGENCE-AVIATION/dp/B0CL7Y4HPJ

6 Artificial Intelligence in Maintenance, Technical Support and Repair Operations

6.1 ROLE OF AI IN PREDICTIVE MAINTENANCE AND CONDITION-BASED (CB) MONITORING FOR VEHICLE SYSTEMS

Ioannis Bakopoulos

Artificial Intelligence (AI) is the ability of machines, such as computers, to search, find, acquire, and process information, build knowledge, and develop reasoning to reach conclusions. AI utilizes Neural Networks (NNs), an interconnected group of nodes. An NN consists of inputs, weights, thresholds, and outputs. This process can be either feed-forward or feed-back (back-propagation), aiming to advance error correction and the learning process of NN. At the core of AI, Deep Learning (DL) unfolds as the most intricate layer, employing multi-layered NN to analyze big datasets and extracting complex patterns and insights that evolve through trial-and-error practices (Read, 2018).

Machine Learning (ML) emerges as a pivotal subset, focusing on algorithms that are trained, supervised, unsupervised, or reinforced, by a data set, to provide insight for either predictions or discoveries. At the top layer, AI is using all the aforementioned learning methods, complemented by other disciplines such as traditional text-to-speech and speech-to-text conversion, robotic motion, and sensors (nowadays known as the Internet of Things, or IoT) but also new ones like Natural Language Processing (NLP) and Visual Recognition (Dorsch, 2018).

In the 21st century, AI is revolutionizing industries with the ability to analyze complex data, predict outcomes, and automate tasks. The impact in the aviation sector is significant in all areas. This chapter will focus on the role of AI in Maintenance and Supply Chain Management (SCM).

6.1.1 INCORPORATING AI INTO AIRCRAFT AND MAINTENANCE – DOCUMENTATION AND SENSOR DATA

The core of Aircraft Maintenance consists of documentation, such as technical manuals, service bulletins, regulations, and aircraft data traditionally collected by inspections or onboard sensors. AI can access, search, find, and correlate a wide

DOI: 10.1201/9781003480891-7

range of maintenance-related information with unprecedented speed and accuracy, facilitating the quick identification and interpretation of data, and reducing downtime during repairs. Thus, AI-driven search capabilities allow maintenance personnel to find specific information, procedures and compliance guidelines rapidly, ensuring maintenance tasks are performed accurately and efficiently, according to the latest standards and regulations.

6.1.1.1 Real-Time Health Monitoring

AI can perform real-time health monitoring of aircraft systems' operational conditions, utilizing IoT sensors and various data sources, to analyze the collected data, identify functional deviations, diagnose potential issues, and even forecast the probable occurrence of component failures.

6.1.1.2 Troubleshooting and Diagnostics

Maintenance personnel can utilize AI to analyze aircraft system outputs, maintenance logs, and sensor data, process complex diagnostic data such as malfunction descriptions or error codes, quickly diagnose issues, and identify potential faults. By comparing these data against the established knowledge base derived from the process of technical manuals, service bulletins, regulations, and maintenance records of already known issues and resolutions, AI can suggest the most likely cause of a problem and recommend corrective actions, reducing diagnostic times and improving accuracy. This process can be facilitated through chatbots where AI can provide step-by-step guidance, offer troubleshooting assistance and answer technical queries on the spot, enhancing the quality and speed of repair work.

6.1.1.3 Real-Time Fault Identification – Virtual Assistants – Automated Diagnostics

AI can analyze verbal directions in post-flight inspections or during flight and access, search, find and correlate a wide range of maintenance-related information with unprecedented speed and accuracy, facilitating precise information retrieval for fault detection/isolation. This information is valuable decision-making assistance for aircraft pilots during flight and for maintenance personnel in post-flight inspections. AI virtual assistants, such as virtual or augmented reality devices, where visual recognition of anomalies and additional visual guidance over the findings can provide maintenance personnel with significant real-time guidance and troubleshooting assistance, improving efficiency, and reducing human error (Fordal et al., 2019).

6.1.1.4 Predictive Maintenance

By analyzing historical maintenance data and real-time performance metrics, AI can predict potential aircraft component or system failures before they occur. This predictive maintenance (PdM) approach allows for the scheduling of maintenance activities at optimal times, reducing significantly unplanned downtimes and extending the operational life of aircraft components.

6.1.1.5 Maintenance Operations Optimization

By collecting and analyzing records and performance data on maintenance operations, AI can identify bottlenecks and inefficiencies within existing workflows,

AI in Maintenance, Technical Support and Repair Operations

recommend process improvements, and optimize workflows. Optimizing these processes ensures that maintenance activities are completed efficiently, minimizing aircraft downtime, improving aircraft availability, and lowering maintenance costs.

6.1.1.6 Quality Assurance and Compliance Monitoring

AI-assisted monitoring tools provide the capability for AI to monitor maintenance activities in real-time, ensuring that they meet quality standards and comply with regulatory requirements. It can provide instant feedback and alert maintenance personnel to deviations from prescribed procedures, helping to maintain quality standards for maintenance activities and regulatory compliance.

6.1.1.7 Continuous Improvement – Risk Management

By collecting and analyzing data from maintenance operations, AI can identify risks, and trends, pinpoint areas for improvement, and suggest changes to optimize repair workflows and develop best practices. This continuous improvement cycle leads to more efficient repair processes, reducing risk, and repair times, lowering costs, and enhancing aircraft reliability over time.

6.1.1.8 Training, Simulations, and Certifications

AI can provide interactive, up-to-date training modules and simulations, tailored for maintenance personnel by analyzing their performance, identifying skill gaps, and recommending personalized learning modules. This ensures that the workforce is well-equipped with the latest knowledge, techniques, and regulations, enhancing the skill set of the workforce and therefore the overall maintenance quality.

6.1.2 SUPPLY CHAIN MANAGEMENT

AI can assist in managing parts inventory, by analyzing historical demand data and extracting various factors, such as demand patterns, lead times, and supplier performance, to quickly identify or even predict parts demand, allowing supply chain managers to optimize inventory levels, reduce stockouts or overstocking, minimize carrying costs, and ensure timely availability of assets.

6.1.2.1 Parts: Identification, Inventory Management, and Procurement

AI monitoring continuously inventory levels, customer location, and operational capacity can automatically generate replenishment orders based on predefined triggers and demand patterns, thus reducing errors in parts procurement, ensuring efficient order processing and fulfillment, optimal stock levels, and timely availability of necessary components, avoiding stockouts, and accelerating repair operations.

6.1.2.2 Supplier Selection and Evaluation

AI can assist in the evaluation and selection of suppliers by analyzing supplier performance metrics, quality ratings, pricing, and delivery reliability, thus enabling supply chain managers to make informed decisions.

6.1.2.3 Warehouse Management

AI can optimize warehouse operations by analyzing various data sets to extract parameters such as order frequency from demand records or optimal storage locations based on product dimensions. Taking into consideration factors such as warehouse storage capacity, storage room dimensions, and distances to cover can lead to reduced picking-up time and unallocated space.

6.1.2.4 Real-Time Tracking and Visibility

AI utilization of IoT sensors, tracking systems, and GPS data provides real-time data analysis and presents optimized visibility into the location and status of all assets throughout the supply chain, improving operational awareness and enabling proactive problem-solving.

6.1.2.5 Logistics and Route Optimization

AI can optimize transportation routes by considering the distance, real-time traffic conditions, fuel consumption, and delivery time windows, leading to improved efficiency, reduced costs, and faster order fulfillment.

6.1.2.6 Supply Chain Optimization

By collecting and analyzing vast amounts of supply chain data from multiple sources, including sales, customer feedback, and external factors, AI can identify bottlenecks and inefficiencies within existing workflows, and provide insights for supply chain optimization, demand shaping and capacity planning.

6.1.2.7 Risk Management in the Supply Chain

AI can analyze various risk factors, such as supplier reliability, geopolitical events, natural disasters and IT vulnerability, to assess potential disruptions in the supply chain and develop contingency plans to mitigate risks.

6.1.3 Conclusion

AI integration can significantly improve maintenance activities and SCM in the aviation sector. The aforementioned examples illustrate how AI can be a valuable tool by leveraging AI-driven insights, predictive analytics, and real-time data processing to enhance decision-making. AI can lead to significant improvements in efficiency, accuracy, reliability, and safety, while at the same time providing a competitive edge to adapt more quickly to rapidly evolving industry dynamics.

REFERENCES

Dorsch, J. (2018, May 30). IIoT and Predictive Maintenance. Semiconductor Engineering online. May 30, 2018. https://semiengineering.com/iiot-and-predictive-maintenance/

Fordal, J.M., Rødseth, H., & Schjølberg, P. (2019). Initiating Industrie 4.0 by Implementing Sensor Management. https://ntnuopen.ntnu.no/ntnuxmlui/bitstream/handle/11250/2649501/Fordal.pdf?sequence=4

Read, B. (2018). Digital Takeover. Royal Aeronautical Society online. https://www.aerosociety.com/news/digital-takeover/ 20 March 2018.

6.2 USE OF AI IN FAULT DETECTION, DIAGNOSIS, AND PROGNOSIS FOR IMPROVED MAINTENANCE EFFICIENCY

Timothy D. Ropp

The desirable attributes of AI for air vehicle maintenance include enhancing the robustness, speed, and accuracy of existing fault detection, diagnosis, and prediction models. Additionally, AI can incorporate human-like attributes of intelligence, such as using local cultural language and making contextual decisions based on ongoing learning and practice through associations. (Copeland, 2024; Cardona et al., 2023). However, AI's human-like attributes must go beyond just novel impersonation. AI must work with assessment, diagnostic, and predictive capabilities on par with the professional intellect of experienced technicians. This is especially critical in context and interpretive analysis, which can be difficult for AI systems (at least for now) and can negatively impact appropriate decision-making.

The study uses the following definitions of fault, diagnosis, and prognosis:

Fault: In terms of aircraft maintenance, a failure or emerging out-of-tolerance condition of a component or system exceeding its type design and operating tolerances, or which could impact safety. Surface irregularities, low operating pressure, high-temperature excursions, or low fluid levels are examples of faults. These can be displayed on screens via the aircraft's health management system or relayed directly to a maintenance ground control center in real time.

Diagnosis: Determining the type and level of severity of a fault condition, isolating it to a specific component or area of defect, and identifying a root cause.

Prognosis: Forecasting a fault's future degradation and failure point, and projecting how much safe operating life or time in service remains. With the latest advancements in AI algorithms, which excel at quickly correlating data, identifying relationships, and making probability forecasts, systems may now also offer relevant maintenance tasks.

6.2.1 Fault Detection

Fault detection and condition monitoring, such as for a gas turbine engine or even an entire airplane, have been a part of aviation flight operations and maintenance planning for many years at varying levels. Engine Condition Monitoring and Aircraft Health Monitoring are two examples. Engine Condition Monitoring is a system of onboard sensors connected to assess parameters, such as sensors on gearbox accessories surrounding an engine core, including rotating component shaft vibrations or temperatures in compartments or along the engine's gas path (Figure 6.2.1).

Small excursions in temperature (temperature delta) or small vibrations beyond preset amplitude limits could indicate part degradation and impending failure, depending on where the anomaly is picked up. Using historical data on that fleet of engines and even during that engine's service life, that condition data is used as early warning indicators of impending failures. This data is gathered in real-time

FIGURE 6.2.1 Remote EGT thermal sensor telemetry testing ambient temperature variations in the Auxiliary Power Unit engine compartment (left) and remote receiver unit (right). Purdue University, AMT-I Center's *Hangar of the Future*, 2024.

using standard sensors and instruments onboard the aircraft itself. The data is transferred to a central facility for analysis and, if needed, adjustments to the maintenance schedule to route that aircraft and its engine into maintenance earlier than planned if needed.

Similarly, many modern aircraft contain central computers and integrated data management units tied to sensors throughout the aircraft. These integrated onboard system fault detection and management systems are known by various names: Boeing's Airplane Health Management Systems (AHMS), Airbus's Skywise Aircraft Health Monitoring System, and Embraer's Aircraft Health Analysis and Diagnosis System are examples of comprehensive IoT and sensor integrated aircraft systems used to detect, diagnose and proactively help predict and head off faults and failures. These comprehensive data architectures incorporate networked sensors throughout the aircraft's major aircraft systems, using cloud-based Wi-Fi and satellite data links to aggregate and provide real-time and future state PdM.

Following the growth of the system health monitoring, Preventive Maintenance (PM) at scheduled times, which incorporates routine inspections for conditions as a failure prevention strategy, remains a prominent method of fault detection in both manufacturing and in-service Maintenance, Repair, and Overhaul (MRO). In addition to human visual inspections, other technologies like die penetrant, ultrasound, and X-rays are employed depending on the critical failure consequences of the part.

These inspections are subject to human factors and shortcomings within these legacy technologies, even when tasks are performed per protocol. Fan disk manufacturing inspections are one example. Missed anomalies – internal material inclusions, voids, or even surface cracks – are always a concern because of their history of being easy to miss. This has resulted in catastrophic failures and accidents. In a modern fan disk manufacturing scenario, automated inspections using ML algorithms learned by a human inspector might, through trial and error, identify anomalies on newly produced disks. However, there are still challenges to overcome, even for machines. Many newly manufactured engine parts are highly reflective. While humans can reposition and view components from discrete angles to obtain a 3D perspective, many existing camera systems view in two dimensions. This would require several

AI in Maintenance, Technical Support and Repair Operations

hundred camera snapshots and still pictures to assemble enough of a mosaic to assess the part in a polished state (Kitov.ai, 2023). Challenges exist for AI's expansion into less controlled environments like MRO in the hangar and line maintenance ramp environments.

Kitov Case Study

Kitov Company has an open software AI inspection system. The Kitov company devised a CAD-based visual inspection system using ML and 3D visioning to identify defects in newly manufactured rotor disks. Using a human expert engineer and a 3D CAD model, the engineer creates an inspection profile using 3D geometry positioning instead of 2D, something more akin to how the human operator would perform. An AI system using a robotic arm mimics this assessment algorithm, learning preferred inspection paths and anomaly characteristics and differentiating between scratches, grease marks or other defects. Kitov reports optimizing disc inspection from 1 hour to 12 minutes (Kitov, 2023).

6.2.2 Fault Diagnosis

Maintenance and servicing tasks occur during an aircraft's service life, where components are often at varying degrees of degradation as part of their planned lifecycle. One component may have significant wear at a point in time compared to its pristine condition when first installed, but it will still be well within the prescribed airworthy and safe operating limits. Component geometry, installation location, and the accumulation of normal dirt and debris during service make inspection and diagnosis a challenge. Many assessments require a level of expert knowledge and experience to determine if a part is faulty and to avoid a false positive.

Similar components may fail at different rates as well. They may leak from unanticipated areas, and break prematurely from unanticipated impact (hail) or heat (lightning strike or leaking bleed air valve). Or the component may have been moved or upgraded based on an FAA directive or engineering change, and its configuration on that particular aircraft is not like the rest of the fleet. These one-off scenarios are regular occurrences requiring high levels of expert cognition and contextual assessment and must be considered. Does the technology help or hinder the technician's and decision-making process?

On the other hand, AI-based technologies do have a use case and have been used in diagnosis, such as Augmented Reality target-based vision as well as laser technologies employed to detect and diagnose anomalies. Augmented Reality and 3D point cloud scanning are used to compare installation alignment and connections. Augmented Reality and other high-definition 3D visioning are used to diagnose turbine engine FADEC cabling misconfigurations and perform routine receiving inspections of housing fin integrity.

Other parts or damage scenarios are more challenging for fault diagnosis and, if not properly assessed, can lead to inappropriate task detection or prognosis.

Is the fluid leak on a pump housing from that particular housing, or is it coming from another source? Is it just a condensation drip? Is the detected fault a problem now, or could it be safely and legally deferred until 20 flight cycles from now when a maintenance base has the replacement parts on hand? Can it run to failure without impacting the safety of flight? These are expert-level assessments based on experience, which a predictive system must be able to replicate with diagnostic precision. In its current state, AI is still prone to shortfalls in light of these contextual nuances, resulting in potential misdiagnosis errors.

Case Study: Standford University's Early AI Model

In the early development of expert AI systems, Standford University developed an expert medical system for diagnosing and treating blood infections. It used 500 rules to operate close to a human specialist for diagnosing various blood infections, and it even exceeded some general practitioners in diagnosis accuracy. But its limitations were soon discovered. In one instance, the program was told that a patient with a gunshot wound was bleeding to death. The program attempted to diagnose a bacterial cause for the bleeding symptoms. An appropriate decision path is made inappropriate by a lack of contextual awareness. Other AI failures included prescribing wildly incorrect drug doses for a patient's weight and incorrectly transposing data (Copeland, 2024).

6.2.3 FAULT PROGNOSIS – PREDICTING FAILURE AND PRESCRIBING FIXES

Many inspections, whether creating a new part or checking an in-service part, involve specialist knowledge and contextual, distinct perceptions and interpretations. Sensor indications may indicate sensor failure, not component failure. Critical among AI assessment and prognosis (predictive life remaining and prescribing a fix) is the ability to:

1. Differentiate between an anomaly and a normal condition.
2. Determine if the anomaly is within tolerance and acceptable.
3. Determine out of tolerance the life remaining and prescribe the corrective action.

REFERENCES

Cardona, M.A., Rodriguez, R.J., & Ishmael, K. (2023). *Artificial Intelligence and the Future of Teaching and Learning: Insights and Recommendations.* Office of Educational Technology.

Copeland, B.J. (Feb.2, 2024). Artificial Intelligence. Also known as: AI. Britannica online. https://www.britannica.com/technology/artificial-intelligence#Overview

Kitov.ai (2023). Aerospace Disk Inspection. Smart AI based visual inspection. https://kitov.ai/?gad_source=1&gclid=EAIaIQobChMIks3JwJGkhAMV0wGtBh0ODA3fEAAYAiAAEgKfMvD_BwE

6.3 AI APPLICATIONS IN REPAIR OPERATIONS AND SPARE PARTS MANAGEMENT
Konstantina Ioannidi

AI emerges as an indispensable tool and technology, revolutionizing engineering applications in repair operations within the transportation industry. Exploiting cutting-edge technologies such as big data analysis, ML, Industry 4.0, and the IoT, PdM becomes a cornerstone for a proactive and efficient maintenance engineering process.

PdM, at its core, tries to predict when a piece of equipment might fail, transforming the transportation sector's approach to maintenance from reactive and preventive to a more "forward-looking" strategy. Through the integration of AI, transportation industries manage to reduce maintenance worktime and operational costs as well as minimize breakdowns, enhancing reliability at the same time.

The basis of PdM lies in the real-time or asynchronous collection, analysis, and assessment of critical data obtained through various sensors and monitoring techniques. Sensors, including stain gauges, crack propagation gauges, hall effect, temperature, humidity, vibration, sock, and radio frequency, among others, play a pivotal role in the effective condition monitoring of motors, surfaces, and electronic equipment in transportation systems. The integration of IoT facilitates the seamless transmission of sensor data to a central computer or cloud for comprehensive analysis, enabling engineers and technicians in the transportation sector to monitor machinery and critical equipment in real time.

ML algorithms, constituting predictive formulas, compare equipment's current behavior with its expected behavior. This proactive approach allows for the early detection of potential faults, enabling timely repairs, replacements, or servicing before failure occurs. The PdM software relies on historical performance data from sensors and hard-copy maintenance records to create an initial dataset for the algorithm. Subsequently, using integrated-based monitoring systems, the software can autonomously identify systems operating outside of predefined conditions, alerting maintenance engineers and technicians to take the necessary actions. The entire process mitigates the risk of catastrophic failure and minimizes maintenance time and costs.

In addition to PdM, AI models prove instrumental in analyzing patterns and predicting the demand for spare parts, aiding in the maintenance of optimal inventory levels and reducing storage costs.

Implementing AI prediction models, complemented by ML algorithms analyzing historical transportation data, contributes to the identification of potential safety risk factors. While real-time monitoring as well as early warning systems are integrated into an AI algorithm, it is crucial to emphasize that AI should complement rather than replace human decision-making. Proper training enables engineers and technicians to validate AI system predictions and make informed decisions based on the AI recommendations, particularly in the realm of repair operations and spare parts management. The collaboration between AI models and human expertise is very important, especially when utilizing cognitive systems to assist maintenance teams in decision-making processes. AI's role extends to helping technicians comprehend troubleshooting and critical information within maintenance documentation through NLP.

Moreover, Research and Development (R&D) personnel can leverage collected data to update and improve the AI algorithm, emphasizing that maintenance expertise and oversight remain essential to ensuring the reliability and safety of transportation systems. The integration of AI in fuel consumption optimization, emissions reduction and overall sustainability underscores its positive environmental impact in repair operations and maintenance at large. Case studies highlight AI's important impact on repair operations and spare parts management across various modes of transportation.

6.3.1 AI in the Aviation Industry (Case Study)

In a publication by Stanton et al. (2023), a review study examines and contrasts the predictive model architectures employed in the aerospace sector. The primary applications center around estimating the remaining useful lifetime (RUL) or end-of-life of the system. This estimation then allows for the interpretation of the operational and/or environmental condition, ultimately leading to the forecast of equipment degradation.

The preferred architectures for RUL estimation include Support Vector Machines, K-Nearest Neighbor, and Random Forest. Moreover, the utilization of Autoencoders and Deep Belief Networks architectures is employed to accomplish the health diagnosis of aircraft engines. The Restricted Boltzmann Machine is employed for aircraft health prediction based on time series sensor data, whereas Convolutional Neural Networks and Long Short-Term Memory Network are utilized to forecast the RUL of aircraft engines using raw time series sensor data. The aviation sector also favors Convolutional Neural Networks for forecasting internal pump leakage in hydraulic systems and problems in pneumatic systems. Recurrent Neural Networks (RNN) are utilized to forecast the spread of bearing defects. Particle filter techniques are employed to estimate the fatigue life of the aircraft body structure and determine the magnitude of flaws in the wings, specifically in relation to Airbus A310 data. The Extended Kalman Filter is used to evaluate the size of fatigue cracks in the airframe and anticipate the future distribution of these cracks.

There are numerous programs worldwide that prioritize Prediction Maintenance. The DAME (Distributed Aircraft Maintenance Environment, 2002) project, which received funding from the UK Engineering and Physical Research Council and involved partners such as Rolls Royce and Universities, had the objective of constructing a grid testbed for distributed diagnosis. Other initiatives strive to enhance problem forecasting and PdM methodologies. The OMAHA (Overall Management Architecture for Health Analysis) project is supported by the German Federal Ministry for Economic Affairs and Energy's Aviation Research Program. Lufthansa Industry successfully created predictive models and a uniform methodology to oversee aircraft conditions in this project. NASA and EU Horizon 2020 have provided funding for university research aimed at developing a comprehensive framework for PdM plans and further advancing this cutting-edge technology.

Not only academic research has been made in this field, but also companies from two decades ago to nowadays are working on tools and techniques based on AI to reduce aircraft and transportation means in general downtime and return to service time. A characteristic example is the SAS Platform (2019), which is an application for intelligent diagnosis related to spare parts optimization for C-130J aircraft, provided by Lockheed Martin and saved 1400 hours of downtime in 3 months for 20 aircraft. As well as the Ascentia4 service by Collins Aerospace in 2020, which applied advanced data management and analytic services to components to reduce 30% delays and cancellations on Boeing 787. Moreover, the F16 Aircraft Structural Integrity Program incorporates big data analysis and Mahalanobis distance metric to assess the structural failures of the fighter aircraft flights during different conditions, heights, velocities, and Angles of Attack. The analysis also focuses on severe cracks on the F-16 wing to evaluate potential risks associated with minimal repairs and prevent future cracks.

6.3.2 AI in Rail Transportation (Case Study)

In the rail industry, AI technology integrated into PdM, plays a critical role in ensuring the safety, reliability, and efficiency of train operations, networks, and infrastructure. AI models and algorithms can identify potential issues by analyzing large amounts of data from sensors installed on tracks and trains. Germany's national railway company uses monitoring systems to predict and prevent costly failures, reduce delays, and advance the reliability of Deutsche Bahn's network (Deutsche Bahn, 2024).

Signaling system malfunctions, early detection of wheel defects and track irregularities are the most common applications of AI-based PdM in rail transportation (Computerword, 2018). Sensor and historical data patterns are processed by ML algorithms and provide rail operators with useful information and graphs to properly handle operation and maintenance needs (Oladimeji et al., 2023).

London Underground implemented this model in 2018. A team of engineers used data analytics to clarify individual subsystem components and map the impact of each parameter on every system. For example, they realized that temperature fluctuations had a significant impact on the failure rate. Moreover, they analyzed historical failure and maintenance data to obtain the probability of failures. Monitoring the systems using sensors, knowing the symptoms before each failure occurrence, and integrating ML models, TfL (Transport for London) predicts whether an event – a failure – is about to happen on a track or on a train subsystem with 75% accuracy.

The objective of the Andromeda project, sponsored by the EU Horizon 2020 program, was to develop intelligent sensor networks integrated with an ML analytics platform (CORDIS, 2020). A PdM solution for railway infrastructure was developed by integrating smart Industrial Internet of Things (IIoT) sensors with AI technology. This system performs real-time monitoring and analysis of vital components' health, while also providing actionable recommendations. The collaboration has developed a certified IIoT autonomous device that can be effortlessly installed in the field, withstands harsh environmental conditions, and does not interfere with ordinary train movement. Furthermore, engineers possess the ability to predict and preemptively address any issues, determining the most opportune timing and specific maintenance requirements.

6.3.3 AI IN THE SHIPPING INDUSTRY (CASE STUDY)

The shipping industry used to rely more on human's decision-making than the other transportation industries. However, ship owners and operation managers aim to improve productivity and reduce maintenance costs by integrating state-of-the-art digital technology. AI and ML aim to achieve marine traffic navigation management using big data from the Automatic Identification System and energy efficiency using sensor data and IoT communication systems (Chin, 2021).

To implement PdM using ML for ship generator engines (GEs), the J. Park 2023 research project gathered and examined data obtained from the ship's alarm monitoring system (AMS) during its operation (Park, 2023). Through engine simulations based on AMS data, both normal and abnormal data were obtained for the ML algorithm. In addition, this study emphasizes the significant ship maintenance costs by proposing a solution supported by the National Research Foundation of Korea. By establishing a criterion value (GCCV) for the condition of the GE, which allows for the intuitive identification of abnormal symptoms based on factors like exhaust gas temperature, engine maintenance records, and AMS data analysis, the research suggests that PdM can effectively mitigate these costs. Implementing regression analysis, a correction factor for the GCCV was derived to adjust for specific engine operating conditions, resulting in the revised GCCV. As a result of this research, an AI algorithm for detecting abnormal symptoms in ship GEs was developed, implemented, and validated.

The primary advantage of implementing PM on ships lies in the capacity to plan corrective maintenance proactively before vessels embark into open waters, where repair tasks become notably more challenging and costlier (Marine Digital, 2024).

REFERENCES

Chin, C.S. (2021). Accessed 14th February 2024 https://tec.ieee.org/newsletter/december-2021/artificial-intelligence-for-maritime-transport

Computerword. (2018). Accessed 7th February 2024, https://www.computerworld.com/article/3427521/how-tfl-is-using-predictive-analytics-to-keep-the-underground-moving.html

CORDIS Horizon. (2020). Accessed 7th February 2024 https://cordis.europa.eu/article/id/418452-system-uses-ai-and-industrial-iot-to-usher-in-new-era-of-predictive-maintenance-in-rail

DAME. (2002). Engineering and Physical Sciences Research Council. Distributed aircraft maintenance environment: DAME. Accessed 1st July 2021 https://gow.epsrc.ukri.org/NGBOViewGrant.aspx?GrantRef=GR/R67668/01

Deutsche Bahn (2024). Accessed 7th February 2024, https://www.deutschebahn.com/en/artificial_intelligence-6935068

Marine Digital. (2024). Accessed 7th February 2024 https://marine-digital.com/article_predictive_maintenance_for_marine_vessels

Oladimeji, D. et al. (2023). Smart Transportation: An Overview of Technologies and Applications. *Sensors*, 23, 3880. https://doi.org/10.3390/s23083880

Park, J. (2023). Accessed 14th February 2024 https://www.sciencedirect.com/science/article/abs/pii/S0360544223026634#:~:text=Predictive%20maintenance%20(PdM)%20is%20a,cost%20and%20effectiveness%20of%20ship

SAS. (2019). Artificial intelligence and IoT analytics keep aircraft operational for crucial missions. SAS UK. Accessed 4th July 2022. https://www.sas.com/en_gb/customers/lockheed-martin.html

Stanton, I. et al. (2023). Predictive Maintenance Analytics and Implementation for Aircraft: Challenges and Opportunities. *Systems Engineering*, 26, 216–237. https://doi.org/10.1002/sys.21651

6.4 AI IN MAINTENANCE AND SUPPLY CHAIN
Konstantina Ioannidi

In today's rapidly evolving landscape, the transportation sector relies heavily on the fusion of AI-driven PdM with SCM. PdM stands as a critical strategy for optimizing asset management. Various techniques, including predictive analysis, DL, and ML, can be employed to achieve this objective. As clarified by the case studies explored previously, AI's indispensable role in predicting equipment failures, optimizing maintenance protocols, and reducing downtime is unmistakable. Integrating sensor measurements, IoT signal transmission, and AI algorithms into transportation operations yields intelligent equipment monitoring, thereby extending the lifespan of transportation assets while minimizing operational disruptions.

Supply chain data analytics integrated into AI algorithms, assist the transportation industry in optimizing workflow and decision-making by utilizing machine-generated data from IoT devices. Such industries adopt smart manufacturing techniques and AI-driven technologies like robotics, autonomous vehicles, and blockchain to automate procurement processes.

Intelligent order tracking and automated supplier selection, facilitated by AI technology, offer a solution to the rising demand for same-day and express delivery services. Innovative AI platforms streamline document coordination among companies, reducing paperwork and improving information flow (Mohsen, 2023). Additionally, AI serves as a dependable tool for data verification in the logistic transport sector, enabling faster approval of transactions and inspections through smart contract management, consequently reducing storage processing times. AI algorithms further enhance fleet tracking, route optimization, and traffic management, strengthening security, transparency, and agility in procurement operations and supplier management.

Supply chain analytics solutions play a crucial role in optimizing workflows and exploiting vast datasets for forecasting and issue identification. By employing statistical data and regression analysis, predictive analysis assists the supply chain industry in mitigating risks and minimizing disruptions. Furthermore, another AI technique adopted by the supply chain industry is descriptive analytics, a form of data mining. This approach involves analyzing large data sets to identify patterns and generate summaries, facilitating informed decision-making in SCM. Additionally, prescriptive analytics serves as a vital tool for supply chain operations, with Supplier Relationship Management software being a popular choice. Supplier Relationship Management enables systematic evaluation of suppliers and vendors, along with strategy development to enhance their performance. Moreover, to enrich the customer experience, feedback data from AI systems undergoes analysis and is integrated into evaluation reports, a process known as cognitive analytics (Farooq, 2024).

The integration of AI in SCM has experienced consistent expansion, showcasing significant potential effects in multiple areas such as demand prediction, distribution, transportation, logistics administration, planning, production, sales, and inventory control. Moreover, AI has the capacity to enhance supply chain performance from Agile and Lean standpoints by enhancing responsiveness and adaptability, minimizing inefficiency, and fostering improved collaboration alongside consumer delight. Nevertheless, it is crucial to emphasize that the incorporation of AI into SCM requires substantial skill and resources and gives rise to ethical concerns about the confidentiality and protection of data. Therefore, it is crucial to engage in strategic decision-making and comprehensive risk assessment before incorporating AI technology into supply chain operations. Publications indicate that the primary usages and benefits of AI integration in SCM, encompassing:

1. Enhancing transparency and mitigating disruptions.
2. Optimizing last-mile delivery.
3. Deploying multiagent systems in order to create collaboration across many organizations in supply chains, thereby facilitating knowledge exchange and efficient coordination.
4. Investigating the application of generative AI in the field of logistics.
5. Strengthening Micro, Small, and Medium Enterprises to increase their ability to withstand and adjust to unpredictable market conditions.

To summarize, supply chain analytics solutions play a crucial role in improving workflow and predicting future outcomes. However, there are notable areas of research that need to be addressed regarding the influence of AI on SCM performance. Challenges encompass the absence of uniformity in the application of AI, complexities in quantifying ROI, and complications in integrating with pre-existing systems. Looking forward, the future of AI in SCM requires broad application, seamless interaction with existing systems, and a heightened focus on addressing ethical and privacy concerns.

REFERENCES

Farooq, M. (2024). Artificial Intelligence in Supply Chain Management: A Comprehensive Review and Framework for Resilience and Sustainability. https://doi.org/10.21203/rs.3.rs-3878218/v1

Mohsen, B.M. (2023). Impact of Artificial Intelligence on Supply Chain Management Performance. *Journal of Service Science and Management*, 16, 44–58. https://doi.org/10.4236/jssm.2023.161004

6.5 AI USES FOR TECHNICAL SUPPORT IN TRANSPORTATION MANAGEMENT – MONITORING OF SYSTEMS: AUTOMATION OF THE RADIO PIREP SUBMISSION PROCESS IN GENERAL AVIATION (GA)

Shantanu Gupta

Automation of the Radio PIREP Submission Process in GA

A real-world application of speech recognition, NLP, and ML technology was demonstrated in the research initiatives to develop a *hands-minimized* methodology for GA pilot report (PIREP) submission (Gupta et al., 2022). The NTSB highlights the need for increased and improved GA PIREP submissions (NTSB, 2017). Researchers developed a 3-tier capability demonstration tool to automatically convert spoken PIREPs to coded PIREPs **without human intervention** (Gupta et al., 2022):

1. **Automatic Speech-to-Text Conversion:** A commercially available speech-to-text transcriber was trained on aviation phonetics and PIREP-specific terminology using a ML-based semi-supervised training approach, to recognize pilot speech and convert the audio PIREPs into text transcriptions for further processing.
2. **Information Extraction from Text:** From the PIREP text transcriptions, specific information (phrases corresponding to the various elements of a PIREP.) was extracted and classified into predetermined categories/labels using an NLP technique called Named Entity Recognition.
3. **Generating PIREP Codes:** With the text transcription, pilot phrases were converted to specific PIREP codes.

REFERENCES

Gupta, S., Deo, M., Johnson, M., & Pitts, B. (2022, January). Automated Speech-to-PIREP: Using Speech Recognition Technology and Natural Language Processing to Generate Weather Reports. *Proceedings of the 22nd Conference on Aviation, Range, and Aerospace Meteorology at the 102nd American Meteorological Society Annual Meeting*, 23–27 January, 2022. https://ams.confex.com/ams/102ANNUAL/meetingapp.cgi/Paper/399371

National Transportation Safety Board [NTSB]. (2017). Improving pilot weather report submission and dissemination to benefit safety in the national airspace system (NTSB/SIR-17/02). National Transportation Safety Board. https://www.ntsb.gov/safety/safety-studies/Pages/DCA15SR001.aspx

6.6 DATA GOVERNANCE: THE BASIS FOR AI IN MAINTENANCE AND REPAIR OPERATIONS
Timothy D. Ropp

Maintaining an air vehicle throughout its lifecycle requires access to and consumption of enormous amounts of data: technical, regulatory/compliance, information, as well as a host of additional instructive information. All of this data must not only be sorted and stored for access but also contextually translated to a maintenance task and applied with precision and timeliness for an aircraft to operate safely, economically, and in regulatory compliance. In this era of Industry 4.0 digitization, mass connectivity, and unprecedented data computing capabilities, modern aircraft maintenance stands on the precipice of a major step-change in the way maintenance is performed. Advances in sensors, digital product definition, ML, rapid simulation, and visualization technologies are becoming ubiquitous operational support tools of the trade. Everyday objects like a torque wrench – together with the aircraft itself – can be networked together to form a digital ecosystem of information. In the work environment, this level of modern connectivity is known as the IIoT (NIST, 2024; Dorsch, 2018). When this digital computing web is combined with computer systems and connected machines whose behaviors and outputs also mimic human learning, reasoning, problem-solving, and decision-making, we arrive at a next-generation framework defined as AI (Copeland, 2024). Another more simplified definition of AI simply says, "automation based on associations…a shift from automating actions to automating decisions" (Cardona et al., 2023).

AI is becoming pervasive across aviation and aerospace as more processes go from mere repetitive, pre-loaded rules-based automation to forward-thinking, dynamic processes controlled by sensor-fed computers that can assess emerging variations, predict faults, and adapt their pre-planned work to adjust. In aerospace manufacturing, for example, changes to machine cut patterns and speeds can be made on-the-fly based on in-process quality/conformity inspections (Webster, 2022). Whether single-engine pistons, small business jets, or a large transport category, aircraft have rapidly evolved to become network-enabled 'Big Data machines' pushing and pulling data high in volume, velocity, and varieties of complex technical information – the classic definition of Big Data (Lutkevich, 2023; Mayer-Schonberger & Cukier, 2013) to the point of operation.

Rapid advances in computing and sensor power have enabled this capability. Concepts like the digital thread, where people can be connected to a product via the internet and advanced visual technologies throughout all parts of its lifecycle (Singh & Willcox, 2018), digital twin, edge computing, machine-to-machine learning (M2M), and AI will become ubiquitous data science tools of the trade. These data science tools are, in fact, imperative to refine and integrate into modern maintenance. Airbus in 2018 installed the Rockwell Collins FOMAX data recording unit onto its fleet, boosting the A320's capability to monitor up to 24,000 parameters, its A330 fleet 40,000, and the newer A350 now generates upwards of 400,000 data points on a single flight (Read, 2018). Such systems can assess emerging failure modes and suggest tailored repair or maintenance interval strategies. Advancing techniques for extracting and analyzing large volumes of high-velocity data, using multiple sources and tools, and compiling it into actionable and useful knowledge is therefore paramount (Brodie, 2019).

These are a few examples of the growing set of data science capabilities in modern aircraft and the challenges MRO providers who maintain them must contend with. When combined with additional computer processing capabilities that also mimic human attributes, like learning, reasoning, perception, problem-solving, and using contextual language to fit the scenario, they materialize into the realm defined as AI (Copeland, 2024).

This chapter looks at the use of AI applied to aviation maintenance from a front-line operations perspective. We consider AI's emerging potential to transform the MRO paradigm critically dependent on expert human assessment and decision-making and still dominated by manual "touch work." In particular, we'll look at AI's role in fault detection, diagnosis, and prognosis – key activities in aircraft MRO. As the industry looks to use AI technology, it is imperative to evaluate the ongoing learning curve and challenges it faces as it evolves from rules-based logic and algorithms used in traditional automation to a controlled trial-and-error "human-style" of learning and expert-level differentiation capabilities.

Perhaps the most tantalizing potential of AI is its potential for efficiency gains through rapid, proactive fault detection, analysis, and decision support in front-line MRO operations, where efficiency and cycle time outcomes are determined – sometimes on a literal minute-to-minute basis. But efficiency and cycle time mean nothing in aviation if they are achieved at the expense of airworthiness and reliability. The performance outcome goal of all maintenance is to restore aircraft components and systems to an airworthy condition. As defined by the Federal Aviation Administration (FAA, 2024) and the International Civil Aviation Organization (ICAO, 2018), airworthiness is achieved when the air vehicle or its components (1) conform to its Type Certificate Data Sheet/Type Design and (2) are in a condition for safe operation. These two outcomes are non-negotiable.

While data security assurance is an entire study in itself, the required use of cloud computing and machine-to-machine data transfer using the internet and the digital thread concept is acknowledged as a critical consideration. Data authentication, access, and version control are of major concern. In the present day, cyber-attacks, viruses, and simple un-commanded data downloads to the wrong person or agency are of significant concern, enough that firewall control and ease but control of permissioned access can impose a time burden on a process, as any person who has experienced two-factor identification could probably attest to.

Trusted identities and data network security assurance are pinch points for AI. Intelligent systems rely on appropriate access to data once (and still) held close to the chest. Using any AI system with its associated MRO prediction models, product lifecycle management and enterprise management software, engineering design tools, and product definition tools, requires organizations to share sensitive and proprietary data with partnering agencies and vendors along the supply and support chain. As noted in one aerospace industry survey on the use of AI and data transparency, "striking the right balance is the difference between digital precision and a missed target" (Gottlieb, 2023).

6.6.1 AI AND MRO BUSINESS METRICS CONSIDERATIONS

It is important to understand the key maintenance processes, drivers, challenges, and performance outcomes of aircraft maintenance. In addition to airworthiness and safety assurance, there are additional quality, reliability, and return on investment business metrics that must be considered for any new system or technology. These include:

- Increased 'positive' wrench time: more purposeful tasks (labor hours) applied to the aircraft.
- Increased First-time fix: reduced errors and rework.
- Increased throughput and work volume.
- Reduced On-The-Job Training and onboarding for new hires. Compressing the learning curve.
- Reduced cycle time.

To achieve these, visibility and understanding of the people, parts, and processes involved and their placement, and how they help achieve those maintenance outcomes are essential. Adopting technology without a clear understanding of it could result in automating broken processes, or overwhelming an unprepared workforce. With these considerations in mind, we will look at three areas MROs contend with daily: fault detection, diagnosis, and prognosis, where AI could potentially help, and the challenges to consider. These inspections are subject to human factors and shortcomings within these legacy technologies, even when tasks are performed per protocol. Fan disk manufacturing inspections are one example. Missed anomalies – internal material inclusions, voids, or even surface cracks – are always a concern because of their history of being easy to miss. This has resulted in catastrophic failures and accidents. The following accident case study demonstrates the essential role of accurate fault detection and the human and technology factors involved.

6.6.2 PREDICTIVE MAINTENANCE AND ITS TOOLS IN MRO

PdM is an extension of CB maintenance. It is similar to, or better stated, a more proactive extension of, the CB maintenance philosophy. Under CB, scheduled, manual inspections are performed to assess the current state of component degradation: is the component's condition within safe, airworthy limits established by the manufacturer or the company's operating specification? If found at the limits of its airworthy service condition or nearly so, a restorative task (typically replenishment or replacement) is performed. Philosophically, the goal is to diagnose faults in progress using what is termed a failure-preventive approach.

While this method is proactive in spirit, it still waits for visible or latent inspection indicators of a component's degraded state. The result is a component potentially moving closer to failure or an unairworthy condition than it is to a normal operating state.

PdM is where AI's data aggregation, analysis and predictive output capabilities, especially in prediction modeling and translation into human-friendly, visualized data using contextual and prescriptive job tasking language are of particular interest to MRO. Using a variety of advanced sensors and historical data trending (linear regression), the PdM strategy attempts to look at and trend early warning anomalies that have been associated with component failure in the past, predict possible failures, and prescribe preemptive maintenance action. As one aerospace manager stated, "don't just give me a thousand lines of sensor data showing me what just broke…Make the system give me three stories of what is being detected and how to address it before it breaks."

There are questions raised in MRO concerning the apparent replacement of what looks like a perfectly good part. Prognostic data from PdM and electing to preemptively replace what historically would be thought of as a still-functioning and economically viable component raises a fair question: why replace a still-functional part?

Part of the answer lies in the emerging power and evolving capabilities of the tools of data science including better prediction modeling. The goal of PdM is to blend both analytical and cognitive technologies to enable accurate, positive, and timely decision-making with a high level of confidence (Dorsch, 2018). An example would be replacing a pump that indicates just a slightly irregular shaft vibration. A leading cause of pump seal failures downstream occurs from early abnormal pump shaft deflection and "wobble," causing repetitive stress cycles around seals. If pump shaft wobble can be sensed at earlier amplitudes by a new generation ultrasonic sensor sooner, the pump and shaft itself could be fixed or changed out before the condition deteriorates to a full-on leak, causing the aircraft to remain on the ground with a delay for troubleshooting, latent diagnosis, and a fix.

Many prediction models use historical data trend analysis along with real-time sensors to make predictions for when specific components of the aircraft are likely to malfunction or require repairs and maintenance (Fordal et al., 2019). Historical data can be retrieved from parts manuals and records, showing how the specific component should be used and when it is likely to need maintenance services. In this kind of modeling, computers run a code using data modeling based on a combination of expected due dates or cycle intervals for inspections or estimated breakage. The aircraft maintenance operators are notified through notifications, alarms, and signals (Fordal et al., 2019). This approach helps prevent unforeseen breakdowns and accidents, as it seeks to maintain and repair the aircraft before it degrades to minimal levels or malfunctions.

REFERENCES

Brodie, M.L. (2019). What is data science? In *Applied Data Science*, pp. 101–130. June 2019. DOI:10.1007/978-3-030-11821-1_8

Cardona, M.A., Rodriguez, R.J., & Ishmael, K. (2023). *Artificial Intelligence and the Future of Teaching and Learning: Insights and Recommendations*. Office of Educational Technology.

Copeland, B.J. (Feb.2, 2024). Artificial Intelligence. Also known as: AI. Britannica online. https://www.britannica.com/technology/artificial-intelligence#Overview

Dorsch, J. (2018, May 30). IIoT and Predictive Maintenance. Semiconductor Engineering online. May 30, 2018. https://semiengineering.com/iiot-and-predictive-maintenance/

Federal Aviation Administration (2024). 14 CFR, Chpt. 1, Sub Chapter C, Subpart F, §21.130. Statement of Conformity. eCFR System. https://www.ecfr.gov/current/title-14/chapter-I/subchapter-C/part-21/subpart-F/section-21.130

Fordal, J.M., Rødseth, H., & Schjølberg, P. (2019). Initiating Industrie 4.0 by Implementing Sensor Management. https://ntnuopen.ntnu.no/ntnuxmlui/bitstream/handle/11250/2649501/Fordal.pdf?sequence=4

Gottlieb, C. (10 July, 2023). AI for MRO Requires Data Transparency. Aviation Week Network. MRO Emerging Technologies. https://aviationweek.com/mro/emerging-technologies/opinion-ai-mro-requires-prioritizing-data-transparency

ICAO. (2018). *International Standards and Recommended Practices. Annex 8. Airworthiness of Aircraft*. Twelfth Edition, pp. 1, Part 1 Definitions. ICAO.

Lutkevich, B. (Feb. 2023). What are the 3V's of Big Data? TechTarget 2023. https://www.techtarget.com/whatis/definition/3Vs

Mayer-Schonberger, V. & Cukier, K. (2013). *Big Data: A Revolution That Will Transform How We Live, Work and Think*. Eamon Dolan/Huffington Millifin Harcourt.

Read, B. (2018). Digital Takeover. Royal Aeronautical Society online. https://www.aerosociety.com/news/digital-takeover/ 20 March 2018.

Ross R, Pillitteri V (2024) Protecting Controlled Unclassified Information in Nonfederal Systems and Organizations. (National Institute of Standards and Technology, Gaithersburg, MD), NIST Special Publication (SP) NIST SP 800-171r3. https://doi.org/10.6028/NIST.SP.800-171r3

Singh, V. & Willcox, K.E. (2018). Engineering Design with Digital Thread. *American Institute of Aeronautics and Astronautics (AIAA) Journal*, 56 (11). November, 2018. Aerospace Research Central. https://arc.aiaa.org/doi/10.2514/1.J057255

6.7 THE PURDUE UNIVERSITY SCHOOL OF AVIATION AND TRANSPORTATION TECHNOLOGY (SATT) CASE STUDY

Timothy D. Ropp

Purdue University (SATT) utilizes emerging technologies in Fault Detection, Diagnosis, and Prognosis for improved maintenance. In addition to trending historical data, PdM also relies on the use of real-time sensor indications to assess the immediate condition of a component's health, including exhaust emissions (Figure 6.7.1) and other data points that can be extrapolated in rapid simulation software and show predicted failure points.

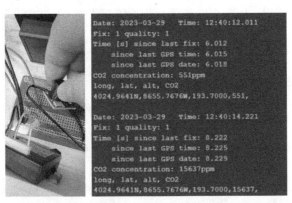

FIGURE 6.7.1 Engine exhaust emissions test data using a CO_2 microsensor on an experimental airborne atmospheric sensing design. Purdue University, AMT-I Center's *Hangar of the Future*, 2023.

AI in Maintenance, Technical Support and Repair Operations

In PdM, data processing is accomplished at the point of operation on the edge of the network. Rather than waiting on traditional data uploading for interpretation by an engineering group, that group's expertise is placed into expert learning algorithms. Sensor-linked machines learn and refine suggested maintenance task decisions from a multitude of networked sensors, forming a more holistic picture of the issue.

6.7.1 Digital Twin Tools in MRO

Digital Twinning is another AI-powered supporting extension of PdM in the presented Purdue University SATT case study. Digital Twins are more than just a model-based replica of a component, engine, or aircraft. Using data aggregation and analysis, Digital Twins usage in PdM is a holistic virtual character capable of real-time multi-system simulation. On-going trend analysis programs, ML algorithms, and aggregated numerical sensor values, combined with 3D representations of the aircraft or component "as flown" state, can produce a powerful AI tool (Hartman & Ropp, 2013). Using the associations and graphics abilities of AI provides the ability to view data overlayed onto 3D models that can show "as manufactured" versus "as flown and maintained" conditions. This, in turn, can help visualize trends across fleets, as well as produce better simulations and assessments for visualizing aging air fleets or even training (Figure 6.7.2).

On a day-to-day level, digital twin datasets are translated into visual context, like an augmented reality or mixed reality overlay of an original component as manufactured versus an in-service part as flown. This provides real-time, expert-level comparisons for damage and repair determination with or without a human being. The Purdue case study uses several tested applications including the use of Augmented Reality to provide step-by-step task instructions (Figure 6.7.3) and networked expert assistance (Figure 6.7.4). In this case, (Figures 6.7.3 and 6.7.4), the expert guides the technician's operations in real-time through their wearable device.

FIGURE 6.7.2 Digital Twin: Virtual engine fault simulation using Unity gaming engine and remote coding and transmit sensor. Purdue University, AMT-I Center's *Hangar of the Future*, 2023.

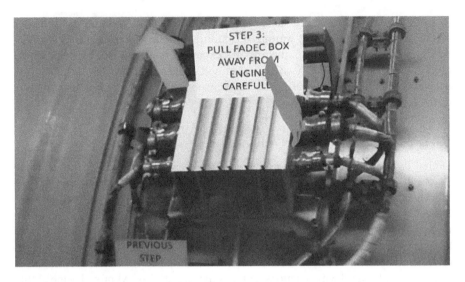

FIGURE 6.7.3 Augmented Reality visual task assistance for removing turbine engine components. Purdue University, AMT-I Center's *Hangar of the Future*, 2023.

FIGURE 6.7.4 Technician using Microsoft Halo Lens with task assistance guidance to perform a flap inspection and service task. Purdue University, AMT-I Center's *Hangar of the Future*, 2023.

REFERENCE

Hartman, N.W. & Ropp, T.D. (2013). *Examining the Use of Model-based Work Instructions in the Aviation Maintenance Environment.* PLM13, tech.purdue.edu/facilities/hangar-of-future.

7 Human-Machine Interaction in Artificial Intelligence in Aviation

7.1 HUMAN FACTOR CONSIDERATIONS IN THE IMPLEMENTATION OF AI IN TRANSPORTATION

Anastasios Plioutsias and Dimitrios Ziakkas

Artificial Intelligence (AI) in transportation has brought about a significant change in how we perceive mobility, promising to improve efficiency, safety, and sustainability. However, the success of these technologies depends heavily on their interaction with human users. Therefore, it is essential to consider the human factor in designing systems that people can use effectively, safely, and comfortably. Human factors are the scientific discipline that studies human interaction and other system elements. In the context of AI in transportation, it is crucial to optimize the human operator's well-being and overall performance by designing systems compatible with their capabilities and limitations.

Human-Centered Design is a framework that prioritizes human perspectives and needs in developing interactive systems. In AI transportation, this approach ensures that technologies are accessible, usable, and beneficial to all users. Case studies show that Human-Centered Design improves user satisfaction and system efficiency, as seen in implementing user-friendly interfaces for public transit navigation apps.

7.1.1 Key Human Factor Considerations

The following key Human Factors are addressed for the AI implementation in transportation research.

7.1.1.1 User Experience (UX) Design

AI transportation systems must provide an intuitive interface and experience that caters to a diverse user base. UI/UX design challenges often arise from the need to accommodate varying levels of technological literacy and physical abilities. Solutions such as adaptive interfaces, personalized settings, and voice-activated controls are essential to improving accessibility and usability for all users.

7.1.1.2 Trust in AI

Establishing trust in AI systems relies heavily on their demonstrated reliability and safety. Transparent communication regarding the system's functionalities, safety protocols, and emergency measures is crucial. Additionally, continuous monitoring and

DOI: 10.1201/9781003480891-8

updating of AI systems contributes to their reliability, which fosters trust among users and operators. Achieving this requires clear and open communication about AI's capabilities and limitations. Ensuring that AI systems make consistent and understandable decisions for human users is crucial. This is necessary to build trust and confidence in these technologies. The use of AI in transportation raises ethical questions about privacy, data security, and decision-making autonomy. Moreover, Artificial has been pivotal in evolving transportation and safety measures (Abduljabbar et al., 2019; Ma et al., 2020). Implementing AI in transportation requires careful thought and consideration of ethical implications, including privacy concerns and potential bias.

7.1.1.3 Training

Training and adaptation are essential for the effective operation of AI transportation systems. The operators and maintenance staff must receive comprehensive training programs that cover technical aspects, operational procedures, and emergency responses. Additionally, AI systems must be designed to adapt to human behaviors and preferences, ensuring a symbiotic relationship between technology and users. Educating users on how AI systems work and how to interact with them effectively is crucial. Organizations and users must be prepared to adapt to changes brought about by AI technologies, including new operational practices and skill requirements.

7.1.2 CHALLENGES AND OPPORTUNITIES

Integrating human factors in AI transportation systems can be challenging due to resistance to change, privacy concerns, and the potential for over-reliance on AI. To address these challenges, organizations should prioritize user education, adopt transparent communication strategies, and develop AI systems that complement human abilities rather than replace them.

7.1.3 CASE STUDY

Autonomous vehicles (AVs) are a prime example of how AI is integrated into transportation. Recent studies on AI in transportation highlight its vast applications, including in AVs and smart cities (Abduljabbar et al., 2019; Nikitas et al., 2020). AI integration in transportation is complicated and emphasizes human-centeredness. The transition between AI and human control, user trust, and ethical decision-making in critical situations are special to AVs. Human considerations must be considered when creating and deploying AV systems.

REFERENCES

Abduljabbar, R.L., Dia, H., Liyanage, S., & Bagloee, S.A. (2019). Applications of artificial intelligence in transport: An overview. *Sustainability*, 11, 189.

Nikitas, A., Michalakopoulou, K., Njoya, E., & Karampatzakis, D. (2020). Artificial intelligence, transport and the smart city: Definitions and dimensions of a new mobility era. *Sustainability*, 12, 2789.

7.2 TRAINING AND SKILL DEVELOPMENT FOR HUMAN OPERATORS WORKING WITH AI SYSTEMS

Anastasios Plioutsias and Dimitrios Ziakkas

The integration of AI in transportation is changing how we handle traffic, operate vehicles, and ensure the safety and efficiency of transport systems. This technological shift requires an evolution in human-machine interaction (HMI), especially in the training and development of human operators. Jaiswal et al. (2021) focus on the necessity of upskilling employees to adapt to AI advancements in multinational corporations. As AI systems handle more complex tasks, understanding the intricacies of HMI becomes essential for safety, efficiency, and progress in transportation technologies (Huang & Rust, 2018).

7.2.1 Understanding AI and Its Role in Transportation

AI technologies, such as machine learning (ML) algorithms, neural networks (NN), and data analytics, are revolutionizing transportation systems. These technologies are responsible for developing AVs, optimizing traffic flow, and enhancing safety features, among other applications. However, even with the advent of these technologies, the role of human operators is still critical. Operators must understand AI functionalities, limitations, and decision-making processes to manage these systems effectively. This dual role of oversight and intervention highlights the need for comprehensive training covering technical knowledge and critical thinking skills (Jaiswal et al., 2021).

7.2.2 Core Skills for Human Operators

The integration of AI in transportation necessitates operators to be well-versed not only in basic operations, monitoring, and troubleshooting but also in understanding AI systems' inputs, outputs, and behaviors. This includes proficiency in cybersecurity to protect data and knowledge of AI's capabilities and limitations. Understanding how AI algorithms function and learn is essential for managing these systems and predicting their behavior in various scenarios. Operators must also excel in decision-making, critical thinking, and communication, skills vital for working effectively in teams and interfacing with stakeholders. They must critically evaluate AI-generated solutions to ensure alignment with safety protocols and operational goals. The work of Wilder, Horvitz, and Kamar (2020) highlights the potential of ML systems to augment human decision-making processes, emphasizing the importance of swift, informed decision-making in high-pressure situations. Accommodating diverse learning styles and addressing ethical considerations and human factors, such as resistance to change and fear of job displacement, are also critical. To address these challenges, adopting agile training methods, blended learning approaches, and industry-academia partnerships can provide the flexibility and resources needed for effective training (Wang et al., 2020).

Successfully integrating AI technologies into transportation requires addressing these multifaceted challenges to ensure safety, efficiency, and human-centered operations. By equipping operators with the necessary skills and fostering a continuous learning environment, transportation systems can leverage AI to enhance operational capabilities while maintaining a strong focus on safety and human factors.

REFERENCES

Huang, M.-H. & Rust, R. (2018). Artificial intelligence in service. *Journal of Service Research*, 21, 155–172.

Jaiswal, A., Arun, C., & Varma, A. (2021). Rebooting employees: Upskilling for artificial intelligence in multinational corporations. *The International Journal of Human Resource Management*, 33, 1179–1208.

Wang, K.-J., Rizqi, D.A., & Nguyen, H.-P. (2020). Skill transfer support model based on deep learning. *Journal of Intelligent Manufacturing*, 32(4), 1129–1146. https://doi.org/10.1007/s10845-020-01606-w

Wilder, B., Horvitz, E., & Kamar, E. (2020). Learning to Complement Humans (Version 1). arXiv. https://doi.org/10.48550/ARXIV.2005.00582

7.2.3 PROMPT ENGINEERING IN GENERATIVE AI APPLICATIONS FOR TRANSPORTATION: AN EMERGING COMPETENCY

Konstantinos Pechlivanis

Generative AI applications have become influential tools in diverse areas, including transportation. Utilization of these technologies, such as large language models, improves efficiency (Ziakkas et al., 2023; Korzynski et al., 2023), safety (Ziakkas & Pechlivanis, 2023), and creativity (Korzynski et al., 2023). However, Ziakkas and Pechlivanis (2023) concluded that crafting skilled and proper prompts is a prerequisite to utilizing generative AI in transportation efficiently. Prompt engineering involves creating input instructions or queries to elicit AI models' desired outputs. Recent studies have emphasized the capacity of prompt engineering to improve the efficiency of generative AI technologies such as ChatGPT (Velásquez-Henao et al., 2023).

Generative AI potential users' intentions for prompting can be summarized as information retrieval, content generation, problem-solving, domain understanding, task assistance, recommendation, explanation, translation, summarization, opinion or review, comparison, prediction or forecasting, storytelling, and guidance or advice. Given the widespread use of AI in industrial applications and ongoing developments, prompt engineering is expected to be an essential competence for future employees.

The selection of prompts in AI engineering is influenced by several factors (Ekin, 2023). Firstly, user intent is crucial to creating prompts that direct AI models to produce relevant solutions. Secondly, model comprehension is essential to understand the capabilities and limitations of AI models (e.g., ChatGPT) to maximize their benefits while minimizing their drawbacks. Thirdly, subject specificity is crucial to ensuring prompts are tailored to the subject discussed. Using domain-specific terminology, context, and illustrations helps AI models understand and respond appropriately to specific tasks. Moreover, clarity and specificity are essential for reducing ambiguity and uncertainty, enabling AI models to provide optimal responses.

Lastly, restrictions, such as answer length or format, are essential for directing AI models to produce outputs that adhere to specified needs. These criteria contribute to effective prompt selection, ensuring AI models generate accurate, relevant, and contextually suitable responses.

Prompt engineering plays a pivotal role in unlocking the full potential of generative AI applications in the transportation industry. A new competency is emerging for employees within the industry (Korzynski et al., 2023; Ziakkas & Pechlivanis, 2023). Organizations and individuals should effectively harness generative AI to enhance efficiency, safety, and innovation in transportation systems by addressing challenges such as specificity, domain adaptation, bias mitigation, and safety considerations through tailored techniques and practical tips. Embracing a collaborative, iterative, and safety-first approach to prompt engineering will pave the way for the responsible integration of AI technologies in shaping the future of transportation.

REFERENCES

Ekin, S. (2023). Prompt Engineering for ChatGPT: A Quick Guide to Techniques, Tips, and Best Practices. *TechRxiv*. Preprint. https://doi.org/10.36227/techrxiv.22683919.v2

Korzynski, P., Mazurek, G., Krzypkowska, P., & Kurasniski, A. (2023). Artificial intelligence prompt engineering as a new digital competence: Analysis of generative AI technologies such as ChatGPT. *Entrepreneurial Business and Economics Review*, 11(3), 25–37. https://doi.org/10.15678/EBER.2023.110302.

Velásquez-Henao, J., Franco-Cardona, C., & Cadavid-Higuita, L. (2023). Prompt engineering: A methodology for optimizing interactions with AI-Language Models in the field of engineering. *DYNA*, 90, 9–17. https://doi.org/10.15446/dyna.v90n230.111700.

Ziakkas, D. & Pechlivanis, K. (2023). Artificial intelligence applications in aviation accident classification: A preliminary exploratory study. *Decision Analytics Journal*, 9, 100358. https://doi.org/10.1016/j.dajour.2023.100358.

Ziakkas, D., Vink, L.-S., Pechlivanis, K., & Flores, A. (2023). *Implementation Guide for Artificial Intelligence in Aviation: A Human-Centric Guide for Practitioners and Organizations*. ISBN 9798863704784

7.3 THE FUTURE OF SIMULATION IN HMI IN AI IN TRANSPORTATION

Anastasios Plioutsias and Dimitrios Ziakkas

Simulation technologies have a crucial role in developing and advancing HMI within AI systems in transportation. These technologies allow for the safe testing and refinement of AI algorithms and HMI designs. They also enable researchers and developers to predict and mitigate potential issues in real-world scenarios. The future of simulation in this field promises even more opportunities for innovation, safety, and efficiency improvements.

Recent studies have explored various aspects of transportation technology, including human-machine interfaces (Rettenmaier et al., 2019), simulation revolutions in matter (Aspuru-Guzik et al., 2018), virtual reality (VR) in manufacturing artificial co-drivers (Lio et al., 2015), the role of VR in building trust in AVs (Morra et al., 2019), traffic models for self-driving cars (Góra & Rüb, 2016), and testing tools for transportation cyber-physical systems (Hou et al., 2016). Today, simulation testbeds are characterized by a combination of VR, augmented reality (AR), and mixed reality (MR) technologies, in addition to traditional computer-based models.

7.3.1 IMPLICATIONS FOR HMI

New trends are emerging in the field of simulation for HMI in transportation. These trends involve using advanced human behavior models and real-time data integration. The concept of digital twins is also gaining popularity, offering a virtual representation of physical transportation systems for real-time monitoring, analysis, and simulation.

These advancements in simulation technologies will have a significant impact on HMI in transportation. They will enhance safety and efficiency, as AI systems can predict and prevent potential accidents more effectively. Furthermore, advanced simulations will facilitate the rapid development and adoption of AVs, providing more dependable and user-friendly transportation options. However, despite these advancements, the simulation of HMI in AI-driven transportation still faces several challenges and limitations. Technical issues such as accurately modeling human behavior, ethical concerns about data privacy, and simulation realism pose significant obstacles.

7.3.2 FUTURE TRENDS

Simulation in Human-Machine Interface (HMI) for AI in transportation is anticipated to make substantial progress in the future. Quantum computing, modeling, and other technologies may change how we create, test, and implement AI in transportation. Finally, global simulation testing standards and frameworks are needed to ensure future transportation technology reliability and safety.

REFERENCES

Aspuru-Guzik, A., Lindh, R., & Reiher, M. (2018). The matter simulation (R)evolution. *ACS Central Science*, 4, 144–152. https://doi.org/10.1021/acscentsci.7b00550

Góra, P. & Rüb, I. (2016). Traffic models for self-driving connected cars. *Transportation Research Procedia*, 14, 2207–2216. https://doi.org/10.1016/j.trpro.2016.05.236

Hou, Y., Zhao, Y., Wagh, A., Zhang, L., Qiao, C., Hulme, K., Wu, C., Sadek, A., & Liu, X. (2016). Simulation-based testing and evaluation tools for transportation cyber-physical systems. *IEEE Transactions on Vehicular Technology*, 65, 1098–1108. https://doi.org/10.1109/TVT.2015.2407614

Lio, M., Biral, F., Bertolazzi, E., Galvani, M., Bosetti, P., Windridge, D., Saroldi, A., & Tango, F. (2015). Artificial co-drivers as a universal enabling technology for future intelligent vehicles and transportation systems. *IEEE Transactions on Intelligent Transportation Systems*, 16, 244–263. https://doi.org/10.1109/TITS.2014.2330199

Morra, L., Lamberti, F., Prattico, F. G., La Rosa, S. E., & Montuschi, P. (2019). Building trust in autonomous vehicles: Role of virtual reality driving simulators in HMI design. *IEEE Transactions on Vehicular Technology*, 68, 9438–9450. https://doi.org/10.1109/TVT.2019.2933601

Rettenmaier, M., Pietsch, M., Schmidtler, J., & Bengler, K. (2019). Passing through the bottleneck the potential of external human-machine interfaces. *2019 IEEE Intelligent Vehicles Symposium (IV)*, 1687–1692. https://doi.org/10.1109/IVS.2019.8814082

7.4 AUTOMATION FRAMEWORK FOR HUMAN DECISION-MAKING AND KEY USES OF AI FOR COGNITIVE INFORMATION PROCESSING

Shantanu Gupta

Decision-making is a crucial cognitive activity that plays a vital role in human behavior and organizational operations. According to the American Psychological Association, decision-making is the cognitive process of choosing a belief or plan of action among multiple possibilities. Decision-making in management involves choosing from different options using available information and analytical assessment Additionally, an adept business decision considers the welfare of all stakeholders, encompassing consumers, staff, investors, and the broader community (Heilig & Scheer, 2024).

7.4.1 Automation in Decision Making

Decision-making is the cornerstone of human cognition, permeating every aspect of our personal and professional lives. However, it is fraught with challenges stemming from cognitive biases, uncertainty from limited information, complexity due to the vast amount of information, limited processing capabilities, and the increasing ambiguity and equivocality of the modern world. Automation offers a promising avenue to address these challenges, leveraging computational power to enhance decision-making processes. Decision automation employs AI, vast amounts of data, and regulations to streamline the decision-making process across various sectors, enhancing productivity and uniformity in decision-making, while minimizing risk and errors in decision outcomes (TIBCO, 2024).

Automated Decision-Making (ADM): ADMs utilize extensive high-quality data from interconnected sources, processing capacity, and algorithms to autonomously make judgments without human involvement. Autonomous Decision-Making systems in self-driving cars and robotics use sensor data; security systems and law enforcement depend on identifying data and criminal histories for recognition, whereas ADMs in public administration use demographic and financial data for anomaly detection.

7.4.2 Types of Decision Automation

There are three main categories of decision automation: *Rule-Based Decision Automation, Data-Driven Decision Automation, and Hybrid Approaches to decision automation.*

Rule-Based Decision Automation (RBDA): RBDA involves the use of predefined logic rules to automate the decision-making processes in various systems and applications. The RBDAs provide consistency, reliability, efficiency, and scalability. Some of the examples of rule-based automation in the transportation industry include:

Traffic Management Systems: Intelligent traffic light control systems use RBDA to optimize traffic flow.

Autonomous Vehicles: AVs rely heavily on RBDA where algorithms make decisions based on traffic laws, sensor data, and environmental conditions for safe navigation, obstacle avoidance, and decision-making.

Data-Driven Decision Automation (DDDA): DDDA uses data analytics, ML, and AI to automate decision-making by evaluating extensive data sets. These algorithms can detect patterns, trends, and insights in extensive information, facilitating precise, efficient, and flexible decision-making. Some of the examples of DDDA in the transportation industry include:

Predictive Maintenance for Fleet Management: In transportation fleets, such as trucks, buses, and trains, ensuring vehicle reliability and minimizing downtime is crucial for operational efficiency and passenger safety.

Demand Forecasting in Public Transport: Data-driven models may analyze historical passenger count data, weather conditions, special events, and other relevant factors to forecast demand, assisting transit authorities to optimize schedules, routes, and fleet allocation, enhancing service quality and operational efficiency.

Hybrid Approaches to Decision Automation: Hybrid approaches to decision automation combine the strengths of rule-based systems and data-driven methodologies to create more adaptive, robust, and intelligent decision-making frameworks. Some examples of hybrid approaches to decision automation in the transportation industry include:

Adaptive (or intelligent) Traffic Control Systems

Dynamic Fleet Management and Routing

Hybrid approaches assist public transit agencies in optimizing routes and schedules to meet passenger demand while minimizing operational costs and maximizing efficiency.

To conclude, Decision-Making in transportation involves choosing from different options using available information and analytical assessment and implementing AI.

REFERENCES

Heilig, T. & Scheer, I. (2024). *Decision Intelligence: Transform Your Team and Organization with AI-Driven Decision-Making*. John Wiley & Sons, Inc. https://learning.oreilly.com/library/view/decisionintelligence/9781394185061/f03.xhtml

TIBCO. (2024). What is decision automation? TIBCO glossary. Retrieved February 10, 2024, from https://www.tibco.com/glossary/what-is-decision-automation

7.5 ADAPTIVE AUTOMATION AND AI IN PREDICTIVE HUMAN PERFORMANCE

Shantanu Gupta

AI and Decision Intelligence (DI) represent the forefront of computational technology, transforming how decisions are made in various domains, from healthcare and finance to transportation, business management, and policymaking. AI technologies have progressed to a level where they are now incorporated into decision support systems for practical applications, significantly influencing decision-making processes. This highlights six roles for knowledge-based systems or intelligent systems: assistant, critic, second opinion, expert consultant, tutor, and automation. The role of AI has evolved to encompass assisting, enhancing, replacing, or automating decision-making, thanks to advancements in approaches like rule-based inference, semantic language analysis, Bayesian networks, similarity measures, and NN. Decision Intelligence, on the other hand, is an interdisciplinary field that combines decision science with AI technologies to enhance decision-making processes. DI leverages data, analytics, and AI models to inform decisions, predict outcomes, and optimize strategies. (Heilig & Scheer, 2024).

7.5.1 Adaptive Automation Framework

- **Expert Systems** are a kind of AI that mimics the decision-making skills of a human expert. An expert system consists of a knowledge base that has domain-specific information and an inference engine that uses logical rules to draw conclusions or choices from the knowledge base. Expert systems are valuable for decision-making when human expertise is limited, costly, or error-prone.
- **Decision Support Systems (DSS)** are computer-based systems that help decision-makers gather and analyze information to solve problems and make decisions. DSS allow decision-makers to analyze potential outcomes, make predictions, and evaluate the consequences of multiple decision options in different situations by combining data and models.
- **Intelligent Decision Support Systems** are designed to integrate specific domain knowledge and exhibit intelligent behaviors like learning and reasoning to aid in decision-making processes. These systems require capabilities such as incorporating both explicit factual information and implicit expert insights, utilizing technologies like ML, NN, and deep learning (DL) for learning and reasoning (Goodfellow et al., 2016).
- **Machine Learning (ML)** for decision-making encompasses the application of algorithms and statistical models to enable computers to improve at tasks with experience, without being explicitly programmed for each specific task. At the core of ML, there are three types of learning processes:
  - Supervised Learning
  - Unsupervised Learning
  - Reinforcement Learning
- **Neural Networks (NN)** are computational models that mimic the structure and function of the human brain, created to identify patterns and tackle intricate issues. They are composed of layers of interconnected nodes known as "neurons," with each connection capable of transmitting a signal between neurons.
- **Natural language processing**: This technology has widespread applications across various industries, significantly improving efficiency, accessibility, and decision-making processes. A few examples of real-world applications of natural language processing include:

- Sentiment Analysis for Brand Monitoring
- Chatbots and Conversational Agents
- Machine Translation
- **Computer Vision**: This technology finds application across various domains, enhancing processes, and creating new opportunities in AVs and facial recognition systems.

To conclude, the integration of Adaptive Automation and AI in predicting human performance offers a revolutionary shift in enhancing operational efficiency and safety across various sectors. By leveraging AI's predictive capabilities, organizations can optimize HMI, pre-emptively identify potential performance bottlenecks, and tailor support to human operators' needs. This not only maximizes productivity but also significantly reduces the likelihood of human error, paving the way for a future where human capabilities are augmented through intelligent automation.

REFERENCES

Goodfellow, I., Bengio, Y., & Courville, A. (2016). III Deep Learning Research. In *Deep Learning*, 106–121. MIT Press.

Heilig, T., & Scheer, I. (2024). *Decision Intelligence: Transform Your Team and Organization with AI-Driven Decision-Making.* John Wiley & Sons, Inc. https://learning.oreilly.com/library/view/decision-intelligence/9781394185061/f03.xhtml

7.6 HUMAN-AI TEAMING BY IMPLEMENTING AI-BASED APPLICATIONS ON CONNECTED ELECTRONIC TRANSPORTATION DEVICES

Shuo Liu

7.6.1 Trajectory-Based Operations

The trajectory of an aircraft in 4D includes the traditional three spatial dimensions, along with time as the fourth dimension. This concept enables Trajectory-Based Operations (TBO), significantly enhancing the capacity to predict and manage a flight's position and timing accurately by improving separation management during different flight stages. With real-time trajectory data transmitted from the aircraft's Flight Management System (FMS) to the ground-based systems via datalink, 4D-TBO facilitates short-term tactical separation via voice communication and long-term strategic separation (e.g., duration over 20 minutes). Consequently, air traffic control (ATC) benefits from the enhanced trajectory predictions, facilitating the issuance of more precise instructions to pilots for an agreed-upon 4D trajectory (4DT).

Transitioning to a full 4D-TBO operational environment will be a gradual process. To assess the performance improvement and the challenges in operational scenarios featuring both 4D-equipped and non-equipped aircraft within the same airspace,

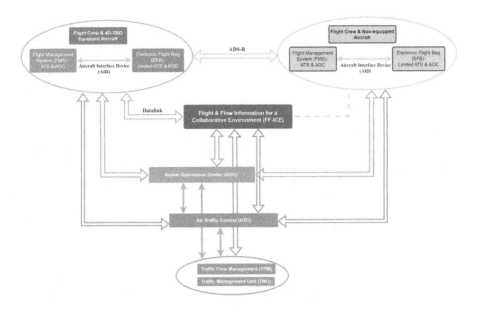

FIGURE 7.6.1 NextGen air traffic control loop presentation.

SESAR, FAA, and CAAC have conducted various flight tests (e.g., SESAR on March 19, 2014, FAA on June 30, 2023, and CAAC on March 20, 2019). These tests, which involve both aircraft types operating simultaneously, simulate the transition period by examining system limitations, information exchange protocols, negotiation of an agreed-upon 4DT, and workload sharing between NextGen 4D-TBO compatible systems and current systems, particularly in trajectory prediction and communications, to determine the collaborative environment needed to implement 4D-TBO.

7.6.2 Air Traffic Control Loop

The current operations are grounded in a labor-intensive and point-to-point system. The NextGen concept is built upon real-time data, common situational awareness, and an approach to collaborative decision-making, with 4D-TBO serving as a key component. Figure 7.6.1 presents the potential use of the NextGen Air Traffic Control loop with 4D-equipped and non-equipped aircraft.

7.6.3 Operation Considerations

7.6.3.1 Trajectory Prediction Errors

The current FMS have the capability to plan and fly efficient trajectories. Despite the enhanced computation capabilities and refined algorithm, there are still certain constraints. Both FMS and ground-based automation systems generate flight trajectories based on wind forecasts. Wind errors lead to trajectory prediction errors.

The FMS's capability to incorporate wind data is limited to waypoints and certain numbers of altitudes at each waypoint. This can pose a challenge, especially en-route, where waypoints are spaced widely apart, making wind predictions less accurate between waypoints. Similar challenges occur during abrupt changes in wind direction and speed, such as wind shear. These limitations can affect the accuracy of predictions for Top of Descent or result in missing the required time of arrival (RTA) (Reynolds et al., 2012).

In the updates provided by the US during the ICAO meeting on utilizing connected aircraft to share reference aircraft trajectory data (ICAO, 2022), it was noted that, "by synchronizing settings and the aircraft state, the same trajectory that would be predicted by the FMS can be provided by the EFB." Implementing new features in the FMS can be challenging, expensive, and time-consuming. Trajectory-based ML applications on a connected device like the Electronic Flight Bag (EFB) provide better adaptability for receiving real-time data from external networks.

7.6.3.2 Trajectory Synthesizer for Ground and Airborne

It was not anticipated that ground automation systems would predict aircraft trajectories as accurately as the FMS. The FMS benefits from access to aircraft-specific parameters such as weight, energy, and performance baseline. In 4D operation, an Extended Projected Profile (EPP) is downlinked via an Automatic Dependent Surveillance-Contract (ADS-C). Using US Oceanic Airspace as an example, meeting the ICAO's proposed minimum separation distance of X (X = 20) nautical miles longitudinally requires an aircraft equipped with ADS-C to transmit position reports every 3.2 minutes (FAA, 2019). Another Mode-S application, Automatic Dependent Surveillance–Broadcast (ADS–B), transmits information about the aircraft's GPS location, altitude, ground speed, and additional information to ground stations and other aircraft every second. In line with the current initiatives by the OpenSky community to utilize automatic ADS-B data, EFB providers will have the opportunity to experiment with different ML models to refine the algorithms for trajectory prediction. Integrating ADS-B and EPP data for trajectory analysis using deep ML models, will greatly enhance the accuracy of predictions and common situational awareness.

7.6.3.3 Trajectory Negotiations

In the transition phase to full 4D operation, the equipage rate of connected aircraft and 4D-TBO capability continues to grow. It's anticipated that the complexity and frequency of trajectory negotiations, as well as the detail of clearance messages for long-term strategic separation (Cardosi & Lennertz, 2022). The current FMS lacks the capability to propose new trajectories, as it is designed primarily to calculate the optimal flight path within the constraints of a predefined flight plan. This means that while the FMS can optimize routes based on factors such as weather, airspace restrictions, and fuel efficiency, it does not actively negotiate or propose alternative trajectories to ATC during flight. Instead, its main function is to ensure efficient navigation and flight execution according to the flight plan already agreed upon or input by the flight crew, with adjustments made for real-time conditions as necessary within the parameters of that plan.

REFERENCES

Cardosi, K. & Lennertz, T. (2022). Flightcrew Human Factors in Message Complexity and Clearance Negotiation. https://rosap.ntl.bts.gov/view/dot/64480

FAA. (July 2019). FAA's Analysis of Costs and Benefits Drove Its Plans to Improve Surveillance in U.S. Oceanic Airspace. https://www.gao.gov/assets/gao-19-532.pdf

ICAO. (September 9, 2022). Use of Connected Aircraft (Ca) for Exchange of Reference Aircraft Trajectory Information [Working Paper]. https://www.icao.int/Meetings/a41/Documents/WP/wp_500_en.pdf

7.7 THE ELECTRONIC FLIGHT BAG CASE STUDY: EXPLORING THE POTENTIAL OF MACHINE LEARNING IN 4D OPERATIONS

Shuo Liu

Significant hidden challenges still exist in terms of both software and hardware, as well as the safety of EFBs when being connected to avionics via an Aircraft Interface Device (AID). Another challenge is that the trajectory algorithm of FMS is certified as a component of the aircraft's type certification process. Therefore, there isn't a separate standard to define the functionalities of the FMS. A third challenge is the delegation between ATC and flight crew, which involves communication, coordination, and decision-making authority.

However, with the enhanced connectivity to both the avionics system and external networks, along with the capability to leverage ML algorithms for processing vast amounts of data, managing repetitive tasks, and recognizing patterns, it comes to the regulatory authorities to establish a certain level of baseline for EFB applications in 4D operation by integrating essential information and tools for non-safety-critical situations. Such standards would enable EFB applications to support synchronized trajectory predictions among the aircraft flight decks (based on configuration and capability), airline ground automation systems, and traffic flow management systems, positioning the flight crew and ATC at the center of the process for reviewing and validating the information on optimized scenarios. Airline flight crews, airline dispatchers, general aviation operators, other airspace users, air traffic controllers, traffic management coordinators, and regulatory bodies involved in certification and operation processes must participate in this project. It is advisable to form a joint working group to examine the scenario comprehensively. Such a collaborative approach would enable the development of more practical recommendations for delineating the roles and responsibilities of trajectory-based ML applications in 4D operations with humans at the center of the design.

8 Ethical and Legal Challenges in the Implementation of Artificial Intelligence

8.1 ETHICAL CONSIDERATIONS IN THE RESPONSIBLE IMPLEMENTATION OF AI IN TRANSPORTATION

8.1.1 AI Ethics in Aviation
Konstantinos Pechlivanis

The ChatGPT release introduced AI applications to the broader public interest and the future potentials AI applications might hold. The aviation ecosystem is already embracing and integrating all the pioneering technological artifacts. Nevertheless, as always, when it comes to the use of science, ethical concerns have emerged. The exponential interest of multiple parties in issuing AI policies denotes the need for ethical guidance. Ethical considerations are critical for maintaining public trust and confidence in the aviation industry, particularly in the context of new and emerging technologies, ensuring fairness and justice, and promoting responsible innovation.

8.1.1.1 Ethics in AI: The Global Landscape

First, AI is such a dynamic concept that it is still defined differently based on discipline perspectives. AI systems can either use symbolic rules or learn a numeric model, and they can also adapt their behavior by analyzing how the environment is affected by their previous actions" (AI HLEG, 2019). Jobin et al. (2019) identified global convergence on at least five ethical principles: transparency, justice and "fairness," non-harmful use, responsibility, and integrity/data protection. At the same time, the European Union (EU), according to the Ethics Guidelines, concluded that a trustworthy AI should be lawful, ethical, and robust throughout its entire lifecycle (AI HLEG, 2019). The essential requirements namely are, namely: *Human agency and oversight; Technical robustness and safety; Privacy and data governance; Transparency; Diversity; Non-discrimination and fairness; Societal and environmental well-being; and Accountability.*

Nonetheless, considerable differences have been highlighted in how these principles are interpreted, why they are essential, what issue, domain, or actors they relate to, and how they should be implemented (Jobin et al., 2019). Even the EU recommendations have been criticized for failing to address modern organizational factors, power, and practices (Larsson, 2020). A plausible hurdle towards establishing an ethical AI consensus is that multiple and diverse stakeholders tend to formulate the ethics of AI in ways that serve their respective priorities. The diversity of the interested parties postpones hard-rule legislative definitions, legal bindings, and policy enforcement (Larsson, 2020). So, is it a stalemate, or can industry and academia move forward?

8.1.1.2 The Ethical Way Forward

Scientific AI research is necessary across multiple disciplines. Clear guidelines and policies should be developed for using AI and other advanced technologies in aviation to ensure that they are used ethically and responsibly (EASA, 2023a, 2023b). Finally, aviation personnel should be provided adequate training and education on using AI and other advanced technologies to ensure they have the skills and knowledge necessary to operate and manage these technologies safely and effectively.

REFERENCES

AI HLEG. (2019). *Ethics Guidelines for Trustworthy* AI. AI HLEG.
EASA. (2023a). *Artificial Intelligence Roadmap 2.0: Human-Centric Approach to AI in Aviation*. EASA
EASA. (2023b). EASA Concept Paper: Guidance for Level 1 & 2 Machine Learning Applications Proposed Issue 02. EASA.
Jobin, A., Ienca, M., & Vayena, E. (2019). The global landscape of AI ethics guidelines. *Nature Machine Intelligence*, 1. https://doi.org/10.1038/s42256-019-0088-2
Larsson, S. (2020). On the governance of artificial intelligence through ethics guidelines. *Asian Journal of Law and Society*, 7, 437–451.

8.1.2 ETHICAL AND LEGAL CONSIDERATIONS IN RESPONSIBLE AI IMPLEMENTATION IN SHIPPING

Michail Diakomihalis

The pursuit and realization of the Principles, Goals and Policies of the General Directorate of the company is the supreme responsibility of the General Manager of the shipping company. The Legal Insurance and Claims Department is responsible for providing legal advice to the company's management for the approval of the company's agreements and contracts with third parties and supporting the Management in all of the company's legal issues. AI implementation has sparked misgivings and controversy, eliciting varied sentiments from people. Uncertainty over the future of workers and society led to concern and disruption in daily life owing to potential fear and effects on the work environment (McClure, 2018). The fundamental basis of the success of the innovative technology lies in the way it is applied and, in the ability, to control it by humans and not put the human factor under its control.

This threat does not appear to be relevant to the shipping industry, as the human presence is vital and irreplaceable. AI will undoubtedly assist and enhance the shipping sector's everyday tasks. Unfortunately, fear hinders the widespread acceptance and use of AI in several human activities, including the shipping sector, despite its acknowledged benefits.

Utilizing AI involves acknowledging issues and surmounting obstacles. Shipping businesses should specifically focus on addressing issues that pertain to them, including:

- The poor quality of the data is an element that often makes it difficult to make appropriate decisions, since the optimal decision is also linked to the quality of the data (Munappy et al., 2019). Lack of dependable and indisputable information pursuing a further improvement in data quality could highlight the lack of reliable information throughout the supply chain, which in turn would also hinder its development. Analytical and qualitative information require and presuppose corresponding data sources of accuracy.
- Job Security concerns; AI implementation is expected to affect the jobs and the evolution of the positions in terms of their subject (McClure, 2018). The industry must keep pace with innovations brought by AI, web applications, sensor technology, and other applications using algorithms.
- The digital transformation of businesses is a process, which requires a certain time for its adoption and implementation and a significant development cost (Heavin & Power, 2018).

As mentioned above, addressing these challenges requires overcoming basic obstacles, which are summarized in the following four categories:

- The integration of available data
- The settlement of issues related to the trust of users and employees/GDPR
- The limitations set in terms of time and necessary energy
- In the lack of innate natural talents, specific skills required

Collaborating with major corporations that utilize AI technology poses a problem for a shipping firm with limited resources and budgets. These businesses aim to verify the anticipated benefit of an AI application before investing resources in the project. Another problem for AI is optimizing the performance of the shipping industry to generate revenues using AI.

REFERENCES

Heavin, C., & Power, D. J. (2018). Challenges for digital transformation-towards a conceptual decision support guide for managers. *Journal of Decision Systems*, 27(sup1), 38–45.

McClure, P. K. (2018). "You're fired," says the robot: the rise of automation in the workplace, technophobes, and fears of unemployment. *Social Science Computer Review*, 36(2), 139–156. https://doi.org/10.1177/0894439317698637.

Munappy, A., Bosch, J., Olsson, H. H., Arpteg, A., & Brinne, B. (2019). Data management challenges for deep learning. In *2019 45th Euromicro Conference on Software Engineering and Advanced Applications (SEAA)* (pp. 140–147). IEEE.

8.2 LEGAL FRAMEWORKS AND CONSIDERATIONS IN THE RESPONSIBLE IMPLEMENTATION OF AI IN TRANSPORTATION

Thanasis Binis, Anastasios Plioutsias, and Dimitrios Ziakkas

AI in transportation has greatly altered our perception of mobility. It offers unparalleled chances to enhance efficiency, safety, and environmental sustainability. It is evident that there is a growing need for strong legal frameworks to guarantee the responsible and ethical application of AI. To achieve this goal, ethical guidelines and legal frameworks are crucial. These frameworks can help harness AI for positive change, keep humans in control, and address the impact of AI on society (Taddeo & Floridi, 2018).

8.2.1 Autonomous Vehicles (AVs)

The regulations regarding the testing and deployment of AVs differ across different regions of the world. The United States of America has a state-centric approach, whereas the Federal Government guides the National Highway Traffic Safety Administration to outline safety standards. In contrast, the EU focuses on a more unified framework for AVs. The EU has issued safety, privacy, and interoperability guidelines across member states. In summary, there are diverse AV testing and deployment regulations across the globe, with the USA adopting a state-centric approach, the EU working towards a harmonized framework, and Asian countries focusing on technological innovation, safety, and security (Bruin, 2016).

8.2.2 Data Protection and Privacy

The EU's General Data Protection Regulation (GDPR) governs AI systems' collection, processing, and storage of personal data. The US California Consumer Privacy Act (CCPA) establishes data privacy standards for AI transportation projects in California and potentially in other states. The GDPR and CCPA emphasize robust data protection standards that affect AI in transportation and individual control over personal data (Pesapane et al., 2018).

8.2.3 Regulatory Compliance and Standardization

Setting global criteria for AI transportation requires international safety and quality requirements. ISO sets these standards, including ISO/SAE 21434 for road vehicle cybersecurity. The International Transport Forum researches and discusses global transportation policies, including AI. The aim is to standardize regulations across different countries. International cooperation is needed to harmonize legislation and address responsibility (Lei et al., 2017). International standards from ISO and International Transport Forum are essential for AI shipping safety and quality.

8.2.4 Conclusion

Legal systems may need help to keep up with the fast-paced development of AI and AV technology, resulting in a regulatory gap that could impede innovation or jeopardize safety. To handle the complicated liability challenges faced by AVs and other AI-driven technology, amended legislation, legal innovation, and new insurance arrangements may be needed.

REFERENCES

Bruin, R. (2016). Autonomous intelligent cars on the European intersection of liability and privacy. *European Journal of Risk Regulation*, 7, 485.

Lei, A., Cruickshank, H., Cao, Y., Asuquo, P., Ogah, C., & Sun, Z. (2017). Blockchain-based dynamic key management for heterogeneous intelligent transportation systems. *IEEE Internet of Things Journal*, 4, 1832–1843. https://doi.org/10.1109/JIOT.2017.2740569.

Pesapane, F., Volonté, C., Codari, M., & Sardanelli, F. (2018). Artificial intelligence as a medical device in radiology: ethical and regulatory issues in Europe and the United States. *Insights into Imaging*, 9, 745–753. https://doi.org/10.1007/s13244-018-0645-y.

Taddeo, M., & Floridi, L. (2018). How AI can be a force for good. *Science*, 361, 751–752. https://doi.org/10.1126/science.aat5991.

8.3 BUILDING BETTER JUST CULTURE IN A WORLD WITH AI
Lea-Sophie Vink

One of the historic criticisms of reporting culture is that it is always focused on the end-user. In Europe, in aviation, the law requires that Air Traffic Controllers (ATCO) and pilots raise safety issues. This means that they are the humans who are investigated, and their actions come under the microscope. While it is true that they hold the liability, more often, the causes of events lie several steps removed from the operators and are caused by much more egregious and risky decision-making, such as from project managers eager to get their changes into operation or engineers who have been sloppy with their tasks. But the front-line operators have to 'clean this up' meaning that they are the safety net for the actions of others which puts them at risk of being blamed or investigated.

This has been the case at lots of organizations. A very busy Air Navigation Service provider in Europe was implementing a new set of electronic flight strips for air traffic controllers and replacing the old paper ones. The project manager was briefed by project staff, including human factors and safety experts, that the usability of the new tools was not good enough to go live. They were overruled by the project manager, who was eager not to look bad and miss out on a bonus. Implementation proceeded poorly, resulting in immediate safety capacity constraints on a busy airport while tools were rectified and enhanced. This led to real costs and lost profits for the airport, which took legal action against the service provider. The ATCOs were the ones investigated for safety occurrences as a result, but the non-tolerated behaviors lay with the project manager. It is for this reason that Just Culture principles and committees must always apply to all persons in an operation and organization, which is a good lesson for those wishing to adopt AI as fast as possible.

8.4 SAFETY MANAGEMENT SYSTEMS CASE STUDY
Anastasios Plioutsias, Dimitrios Ziakkas, and Konstantinos Pechlivanis

Integrating AI into Safety Management Systems (SMS) significantly advances operational efficiency and risk management across various industries, such as aviation, healthcare, and industrial safety. With the ability to process vast amounts of data and predict potential hazards, AI has the potential to reduce accidents and save lives significantly. However, integrating AI with SMS raises ethical and legal issues. To properly use AI to increase safety while respecting individual rights and legal norms, these obstacles must be addressed.

8.4.1 Case Study Overview

The aviation SMS selected case study demonstrates how the airline SMS adopted an AI-driven anomaly detection system. This AI application aims to increase SMS predictiveness via early detection of safety concerns and operational inefficiencies. The proposed AI model analyzes historical and real-time aircraft sensor and system data to identify safety hazards. This enables pre-emptive actions to be taken to mitigate risks.

8.4.1.1 Context and Objectives
The proposed areas of implementation of AI in SMS are the following:

- Predictive Safety Analysis: Use AI to analyze historical and real-time data to predict potential safety incidents before they occur.
- Operational Efficiency: Identify inefficiencies through AI analysis to support decision-making processes and improve safety.
- Customization and Scalability: Create an AI model that can be customized for different aircraft types and scaled across various airlines and operational contexts.

8.4.1.2 Implementation Details
Focusing on the Data Collection, the Purdue – Coventry Universities and HF Horizon research team integrated a range of data sources, including flight data recorders, maintenance logs, and weather reports, to provide the necessary data for the proposed AI model. Regarding the Model Development, data scientists and aviation safety experts collaborated to create and train an AI model capable of identifying complex patterns that could indicate safety risks. Finally, respecting the Ethical and Legal Considerations, the research team established protocols for data privacy, implemented bias mitigation strategies, and adhered to aviation-specific regulations and international safety standards.

8.4.1.3 Challenges and Opportunities

The implementation of AI must assure the accuracy and privacy of data and overcome the resistance to the adoption of Artificial Intelligence (AI) among stakeholders. Additionally, it establishes a precedent for the incorporation of AI in high-stakes industries, improves safety outcomes, and enhances the predictive capabilities of SMS.

REFERENCE

Ziakkas, D., Vink, L.-S., Pechlivanis, K., & Flores, A. (2023). *Implementation Guide for Artificial Intelligence in Aviation: A Human-Centric Guide for Practitioners and Organizations*. ISBN 9798863704784

9 Best Practices for the Implementation of Artificial Intelligence

9.1 COMMON CHALLENGES AND LIMITATIONS OF IMPLEMENTING AI IN TRANSPORTATION

Konstantinos Pechlivanis

The future of intelligent transportation systems and autonomous vehicles is being shaped by advancements in Artificial Intelligence (AI). Integrating AI into transportation poses numerous hurdles and constraints, despite its potential advantages, such as the following:

- **Data Quality and Availability**: Fragmented, inconsistent, or outdated data can hinder the effective training of AI models (EASA, 2023).
- **AI Integration into Current Infrastructure and Processes**: Transportation networks are typically defined by outdated systems, various technologies, and intricate regulatory structures.
- **Ethics and Governance**: The integration of AI into transportation faces ethical and regulatory challenges, including fairness, transparency, and accountability in decision-making processes. Establishing rules, guidance, and governance for AI ethics is a challenging task with no consensus among experts.
- **User Acceptance and Skills' Gap**: Public trust and acceptability (Endsley, 2023) are essential for the widespread use of AI technology such as autonomous vehicles, predictive maintenance systems, and route optimization algorithms.

By innovating, collaborating, and making strategic investments, we can overcome limitations and maximize the potential of AI to transform transportation systems for a safer, more efficient, and sustainable future.

REFERENCES

EASA. (2023). EASA Concept Paper: Guidance for Level 1 & 2 Machine Learning Applications Proposed Issue 02.

Endsley, M. R. (2023). Ironies of artificial intelligence. *Ergonomics*. https://doi.org/10.1080/00140139.2023.2243404

9.2 BEST PRACTICES FOR SUCCESSFUL IMPLEMENTATION OF AI IN TRANSPORTATION, INCLUDING RISK ASSESSMENT, CHANGE MANAGEMENT, AND COLLABORATION

Lea-Sophie Vink

9.2.1 Introduction

The integration of AI into transportation systems promises to revolutionize how we move goods and people. From autonomous vehicles to predictive maintenance, AI technologies offer the potential to enhance efficiency, safety, and sustainability. However, the successful implementation of AI requires careful planning, risk assessment, change management, and collaboration.

9.2.1.1 Risk Assessment

Identifying Potential Risks: Implementing AI in transportation involves technical, regulatory, and ethical risks. Technical risks include data security and system reliability, while regulatory risks encompass compliance with existing and forthcoming legislation. Ethical considerations, such as privacy and job displacement, must also be addressed.

Mitigating Strategies: Effective risk mitigation begins with comprehensive risk assessment exercises, involving stakeholders across the organization. Developing robust data protection measures, ensuring system redundancy, and fostering transparency can address many of these concerns. Additionally, engaging with regulatory bodies early in the implementation process can help navigate the complex regulatory landscape.

9.2.1.2 Change Management

Preparing the Workforce: The advent of AI technologies may necessitate new skill sets, potentially leading to workforce displacement. Proactive change management strategies, including upskilling and reskilling programs, can prepare employees for the transition, mitigating resistance and fostering a culture of innovation.

Phased Integration Approach: Implementing AI technologies through a phased approach allows for the gradual integration of AI systems, enabling the workforce to adapt to new processes and technologies. This method also facilitates the identification and resolution of issues in the early stages of implementation.

9.2.1.3 Collaboration

Industry Partnerships: Collaboration with technology providers, academic institutions, and industry consortiums can accelerate the development and integration of AI technologies in transportation.

Regulatory Compliance: Engaging with regulatory bodies and participating in the development of industry standards can ensure compliance and influence the regulatory environment. Collaboration with policymakers can also promote the development of supportive legislation and regulatory frameworks.

Best Practices for the Implementation of AI

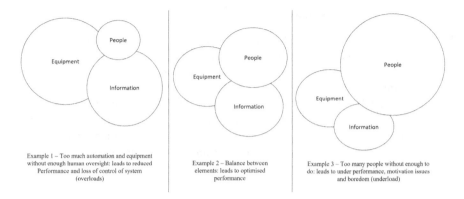

FIGURE 9.2.1 Examples of what goes wrong with the balance between elements (Ziakkas et al., 2023).

One of the most crucial elements to understand when considering the implementation of AI within any safety, high-performing, or high-resilience operation is that, at its core, AI is not a solution for the entire operation – at least in the next 20 or so years. This means that there must be a 'concept of operations' where the idea of humans interacting with information using equipment or tools must also be kept in balance. As Figure 9.2.1 shows, if there are too many humans and not enough tasks, information, or tools, then the human operators can become underloaded, which reduces performance. If humans are not able to comprehend the information or use the tools, this leads to cognitive overload and a loss of performance. Both outcomes are not sustainable, and human factors engineers strive to maintain a balance.

This chapter illustrates the importance of structuring conversations and system design around an operational idea centered on 'automation' with AI tools as a component rather than presuming that AI will completely replace functions. We can gradually integrate AI technologies in a controlled manner to ensure that the changes result in improved performance in areas such as safety, traffic management, and efficiency.

9.2.2 Automation as the Concept of Operations

As shown in Figure 9.2.1, the idea is to find a balance between humans and machines. As the number of tasks increases, such as monitoring increasingly sophisticated systems, humans struggle to maintain that balance. Increasing automation, or the number of tasks that do not require complex decision-making, is the main concept of operations that operations utilize as 'automation as a solution' to the complexity of task management. But Automation has become the solution for concepts of operation across the entire transportation industry. At the heart of understanding how humans remain in the loop with automation is the need to understand how it mediates the relationship between people, information, and equipment (Figure 9.2.1).

9.2.3 THE LEVELS OF AUTOMATION

If there is too much automation, then the success of humans in controlling, monitoring, and operating equipment is undermined. The same is true if there is too little support (Wickens, 2015). Following Wickens & Holland further, we discover that there are four primary ways that information must be processed by the brain, first introduced by Parasuraman and Wickens (2008): Information Acquisition; Information Analysis; Decision Making and Selection; Action Implementation. The extent to which AI supports the four categories will also help guide all subsequent discussions concerning legislation, the responsibility of the operator, and the necessary safety assurance that will be required.

9.2.4 CONCLUSIONS

For every change to the system, operators should understand what levels of automation they have and where they are going. If AI is selected as a tool to increase automation, key questions should always be asked (Ziakkas et al., 2023). Fundamentally, operations should be aware of the levels of automation in their concept of operation, and then test how technologies, people and information may be affected for future concepts.

REFERENCES

Parasuraman, R., & Wickens, C. D. (2008). Humans: still vital after all these years of automation. *Human Factors: The Journal of the Human Factors and Ergonomics Society*, 50(3), 511–520. https://doi.org/10.1518/001872008x312198

Wickens, C. D. (2015, September 2). *Engineering Psychology and Human Performance*. Taylor & Francis. https://www.taylorfrancis.com/books/mono/10.4324/9781315665177/engineering-psychology-human-performance-christopher-wickens-justin-hollands-simon-banbury-raja-parasuraman

Ziakkas, D., Vink, L.-S., Pechlivanis, K., & Flores, A. D. (2023). Implementation Guide for Artificial Intelligence in Aviation: *A Human-Centric Guide for Practitioners and Organizations*. https://www.amazon.com/IMPLEMENTATION-GUIDE-ARTIFICIAL-INTELLIGENCE-AVIATION/dp/B0CL7Y4HPJ

9.3 SIMILARITIES AND DIFFERENCES BETWEEN THE APPLICATION OF AI IN AUTOMOTIVE AND AVIATION INDUSTRIES

Spyridon Chazapis and Dimitrios Ziakkas

At a conceptual level, an automation system is characterized by two main dimensions (Norman, 1988): the "degree of autonomy and the degree of authority." The two combined can create the *illusion of perceived animacy* (the systems seem to have "a mind of its own") – a misconception that is only expected to be amplified with the advent of advanced AI systems. Several distinct levels of authority/delegation are understood to exist, with the most influential model describing a total of ten (Table 9.3.1).

TABLE 9.3.1
Levels of Automation and Corresponding Human/Computer Role in Each (based on Endsley & Kaber, 1999)

	Roles			
Level of Automation	Monitoring	Generating	Selecting	Implementing
1. Manual control	Human	Human	Human	Human
2. Action support	Human/computer	Human	Human	Human/computer
3. Batch processing	Human/computer	Human	Human	Computer
4. Shared control	Human/computer	Human/computer	Human	Human/computer
5. Decision support	Human/computer	Human/computer	Human	Computer
6. Blended decision making	Human/computer	Human/computer	Human/computer	Computer
7. Rigid system	Human/computer	Computer	Human	Computer
8. Automated decision making	Human/computer	Human/computer	Computer	Computer
9. Supervisory control	Human/computer	Computer	Computer	Computer
10. Full automation	Computer	Computer	Computer	Computer

With these basic theoretical elements of automation in mind, a pattern starts to emerge concerning how aircraft and automobile automation, at their current iterations, fit into the overall scheme of "how advanced is it?". Furthermore, similarities (and significant differences) between the two can be identified, as can the potential effects of introducing AI to the mix. Current-generation automobiles come with a variety of safety-related systems (some of which are mandated by law in several markets). The identified categories are:

Typical Advanced Driver-Assistance Systems: *Collision avoidance systems; Adaptive Cruise Control; Traffic sign recognition.*

Similar systems have been installed in aircraft for a while – although aircraft navigation is realized in 3D space, the problem itself is more constrained, and acceptable levels of functionality can be achieved using less developed technologies.

Typical Safety-Related Applications of Automation in Current-Generation Airliners: *Ground Proximity Warning and Avoidance Systems; Flight Path Optimization; Automatic Traffic Collision Avoidance Systems.*

The latest commercial aircraft include systems capable of executing avoidance maneuvering with operator oversight (Level 9 in Endsley's and Kaber's taxonomies).

To conclude, safety aside, several automated systems are installed in both automobiles and aircraft, with applications ranging from convenience functions (environmental controls, passenger entertainment systems, semi- or fully autonomous parking, and so forth), streamlining of maintenance tracking, and identification/management of technical failures, and others. Beyond the flight deck/automobile cabin, automation can improve manufacturing processes, optimize logistics, and rationalize supply chains – all of these can eventually lead to significant economies of scale.

REFERENCES

Endsley, M. R., & Kaber, D. B. (1999). Level of automation effects on performance, situation awareness and workload in a dynamic control task. *Ergonomics*, 42(3), 462–492.

Norman, D. A. (1988). *The Psychology of Everyday Things*. Basic Books.

9.4 IMPLEMENTATION OF REQUIREMENTS SYSTEMS – START NOW OR MISS THE CHANCE

Anastasios Plioutsias and Dimitrios Ziakkas

Integrating AI into transportation can transform how we approach mobility, enhance operational efficiencies, and improve safety. A requirements system organizes project stakeholders' needs. AI transportation requires functional/non-functional requirements, industry norms, and ethics. The AI project lifecycle relies on it for success. Safety, reliability, and interoperability of AI transportation systems require ISO/TS compliance.

To effectively identify and satisfy all parties' needs and expectations, stakeholders must be engaged early and frequently throughout the AI project's lifetime. This process should use interviews, workshops, and surveys to get diverse feedback. Clear and succinct documentation of these requirements is essential for project team communication. Project deliverables must be regularly validated and verified against these defined requirements to meet stakeholder expectations and industry standards. Implementing these strategies can boost project success rates, meet or exceed stakeholder expectations, and contribute to organizational goals.

The presented case studies in this book show successful AI implementations in transportation through practical requirements systems. In Chapter 10, we summarize key points and offer a perspective on requirements systems in the evolving landscape of AI in transportation.

10 Future Directions and Emerging Trends in Artificial Intelligence in Transportation

10.1 FUTURE DIRECTIONS IN AI TECHNOLOGIES

Anastasios Plioutsias, Dimitrios Ziakkas, and Konstantinos Pechlivanis

The integration of Artificial Intelligence (AI) in the transportation sector marks a significant revolution, enhancing how mobility and logistics operations are conducted and perceived. These innovations not only promise to elevate the efficiency, safety, and sustainability of transportation but also indicate a shift towards more integrated and Human-/User-centric transportation experiences. At the heart of this transformation are autonomous vehicles (AVs), which embody the convergence of AI technology and transportation. Beyond vehicle autonomy, AI's role extends to infrastructure management, where it offers solutions to reduce congestion and predict maintenance needs, potentially leading to smarter, more adaptable urban environments that respond dynamically to transportation demands (Ziakkas, 2023). The widespread implementation of AI in transportation presents significant challenges, especially concerning data privacy, ethical practices, and the necessity for strong legal frameworks to prevent misuse. AI's potential to improve transportation safety using predictive analytics and its role in promoting environmental sustainability by reducing emissions and supporting electric vehicles highlight its importance. Achieving these advantages depends on interdisciplinary cooperation, meticulous regulation, and a strong dedication to ethical principles. This ensures that technological progress is in line with wider societal beliefs and has a positive impact on the future of transportation.

REFERENCE

Ziakkas, D., Vink, L.-S., Pechlivanis, K., & Flores, A. (2023). *Implementation Guide for Artificial Intelligence in Aviation: A Human-Centric Guide for Practitioners and Organizations*. ISBN 9798863704784

10.2 EMERGING TRENDS IN AI TECHNOLOGIES

Anastasios Plioutsias, Dimitrios Ziakkas, and Konstantinos Pechlivanis

The introduction of AVs stands as a hallmark of AI's transformative potential within this sector. AVs leverage advanced sensors, machine learning algorithms, and real-time data analytics to navigate roads with minimal human intervention. This technology aims to reduce the number of traffic accidents attributed to human error, streamline traffic flow, and alleviate congestion, marking a significant leap towards safer and more efficient urban mobility (Ziakkas, 2023).

Beyond vehicle autonomy and maintenance, AI's application in traffic management transforms urban travel dynamics. By analyzing vast datasets, including traffic patterns and weather conditions, AI systems can dynamically adjust traffic signals and recommend optimal routes to drivers, effectively mitigating congestion and reducing travel times. Through real-time data and individual preferences, AI facilitates tailored route planning and public transport adjustments, promising improved service delivery and passenger satisfaction (Ziakkas, 2023).

However, data privacy, ethics, and technological reliability challenges are present. Despite these challenges, quantum computing, 5G connectivity, and machine learning techniques seem promising for AI-powered vehicles. Addressing the challenges associated with AI's integration into transportation is essential, but the potential benefits suggest a bright future for intelligent mobility systems (Ziakkas, 2023).

REFERENCE

Ziakkas, D., Vink, L.-S., Pechlivanis, K., & Flores, A. (2023). *Implementation Guide for Artificial Intelligence in Aviation: A Human-Centric Guide for Practitioners and Organizations.* ISBN 9798863704784

10.3 PREDICTIONS AND PROSPECTS FOR THE FUTURE OF AI IN TRANSPORTATION

Anastasios Plioutsias, Dimitrios Ziakkas, and Aspasia Argyrou

AI is significantly transforming the transportation sector, enhancing public transit systems' efficiency, and leading the development of AVs. With the progression of the Internet of Things, machine learning, and data analytics, AI's application in transportation has become more sophisticated, enabling more precise traffic predictions, improved route planning, and enhanced operational efficiency. Interconnectedness will improve traffic management and vehicle safety, making transportation more efficient, safer, and Human User-Centric (Ziakkas, 2023).

The realm of AVs is particularly indicative of AI's transformative potential in transportation. This expansion signifies a shift toward more integrated transportation networks, where AI-driven analytics and predictive modeling play pivotal roles in optimizing public transit operations and schedules in real-time. Furthermore, as AVs incorporate advanced safety features, communication systems, and environmental considerations, they underscore AI's central role in advancing future transportation networks, promising a landscape marked by significantly reduced traffic accidents and improved overall safety. However, ethical constraints, privacy concerns, and the need for thorough legal frameworks make it challenging to fully realize AI's potential in transportation. As AI continues to transform the transportation industry, adaptable AI development is critical to capturing its full potential while negotiating the complexity of innovation (Ziakkas et al., 2023).

REFERENCE

Ziakkas, D., Vink, L.-S., Pechlivanis, K., & Flores, A. (2023). *Implementation Guide for Artificial Intelligence in Aviation: A Human-Centric Guide for Practitioners and Organizations*. ISBN 9798863704784

10.4 BIG DATA AND SAFETY INTELLIGENCE POSSIBILITIES

Dimitrios Ziakkas, Anastasios Plioutsias, and Dimitra Eirini Synodinou

Integrating big data analytics into transportation systems has significantly enhanced safety intelligence by utilizing predictive analytics to analyze historical data, real-time information, and advanced algorithms. This approach allows for the proactive identification of potential safety hazards across various transportation modes, revolutionizing risk assessment and mitigation strategies. By leveraging data from traffic sensors, GPS devices, weather stations, vehicle telematics, and cameras, transportation agencies and companies can identify risk factors that contribute to accidents before they occur (Ziakkas, 2023).

The integration of predictive analytics into transportation management systems provides real-time insights that support quick decision-making, enhance traffic management, route planning, and resource allocation during adverse conditions. By promoting data-driven decision-making, predictive analytics enables transportation stakeholders to prioritize safety initiatives effectively and assess the impact of their interventions. The continuous improvement and adaptation facilitated by predictive analytics are critical for maintaining resilient transportation systems that effectively mitigate safety risks. Transportation stakeholders can improve their safety culture by reviewing past interventions and updating predictive models with real-time data. To conclude, predictive analytics will improve transportation safety by reducing risks, preventing accidents, and saving lives through more informed and proactive safety management practices as technology advances and data sources increase (Ziakkas, 2023).

REFERENCE

Ziakkas, D., Vink, L.-S., Pechlivanis, K., & Flores, A. (2023). *Implementation Guide for Artificial Intelligence in Aviation: A Human-Centric Guide for Practitioners and Organizations*. ISBN 9798863704784

10.5 THE ADVANCED AIR MOBILITY CASE STUDY: REGULATORY HURDLES AND COMPLIANCE CHALLENGES

Dimitrios Ziakkas, Anastasios Plioutsias, and Yolanda Vaiopoulou

Advanced Air Mobility (AAM) represents a significant leap forward in transportation, utilizing AI to enhance safety, efficiency, and responsiveness in aerial systems. Autonomous drones, powered by machine learning algorithms, exemplify the potential of AAM by operating safely and efficiently within complex networks. However, AAM technology has outpaced regulatory frameworks, providing a major issue for operators and stakeholders. Fragmented regulations cause discrepancies in safety standards, certification processes, and operating needs among countries. This makes it harder to integrate AAM technology into aviation ecosystems, maintain safety standards, and deploy new aircraft and technologies commercially. Certification and integration with airspace infrastructure are AAM's biggest challenges. Advanced technologies like distributed electric propulsion and autonomous flight systems require new safety, risk, and certification norms for AAM vehicles.

To navigate these regulatory challenges effectively, a collaborative approach involving regulators, industry stakeholders, academics, and the public is paramount. Moreover, harmonizing international standards and promoting collaboration across borders can enhance safety and interoperability, facilitating the global adoption of AAM. Addressing infrastructure needs, such as the development of vertiports and charging stations, also requires joint efforts from the public and private sectors. Ultimately, advancing the AAM industry will demand streamlined certification processes, integrated airspace management, and a concerted effort to align regulatory frameworks with the innovative potential of AAM technologies, paving the way for their safe and sustainable deployment.

10.6 CONCLUSIONS
Dimitrios Ziakkas, Anastasios Plioutsias, and Konstantinos Pechlivanis

The advent of self-driving cars and electric Vertical Take-Off and Landing (eVTOL) aircrafts, powered by sophisticated AI algorithms, is poised to fundamentally transform urban mobility, heralding a future marked by increased safety, environmental sustainability, and economic efficiency. These technological advancements promise to revolutionize traffic management, predictive maintenance, and the customization of transportation services, offering a glimpse into an era where daily commutes are not only safer but also more attuned to the individual needs of users.

However, realizing the full potential of these advancements is not without its challenges. AI use in transportation is hindered by technological issues, ethical concerns, the need for extensive regulatory frameworks, and infrastructure improvements. To make the transition to AI-driven transportation systems seamless and advantageous for all parties, researchers, regulators, industry stakeholders, and the public must work together. A coordinated effort is needed to overcome the barriers to integrating AI into our transportation networks and ensure that the transition promotes innovation and follows ethical and regulatory standards.

The scope of this book is to foster a culture of innovation, collaboration, and supportive regulatory practices for integrating AI into transportation, transforming it into a force for social and environmental progress. Ultimately, embracing AI's potential in transportation means working towards a future where mobility enhances the quality of life for society at large, embodying our collective aspirations for a safer, cleaner, and more interconnected world.

Index

academics xviii, xxiii, xxv, 28, 29, 49, 55, 102, 136, 144
accident investigation i
accuracy 10, 13, 35, 37, 38, 40, 41, 42, 43, 72, 73, 89, 94, 96, 97, 100, 103, 126, 130, 134
administrators 27
AI i, xviii, xix, xx, xxi, xxii, xxiv, 1–23, 26–31, 33–39, 45–47, 49–65, 67–97, 99–111, 113, 115–124, 128–144
air i, 8, 16, 20, 23, 29, 30, 32, 42, 45, 46, 48, 53, 55, 57, 58, 62, 64, 65, 68, 74, 85, 87, 88, 91, 95, 97, 99, 108, 109, 113, 124, 125, 127, 132
Air Navigation Service provider 65, 66, 132
air traffic control 8, 29, 45, 46, 53, 124, 125
Airbus xxiv, 41, 98, 102, 108
aircraft xxiv, 15–17, 20, 23, 28, 31, 32, 39, 40, 41, 42, 43, 44, 45, 46, 53, 54, 58, 59, 65, 68, 71, 75, 93, 94, 95, 97, 98, 99, 102, 103, 104, 105, 108, 109, 110, 111, 113, 124, 125, 126, 127, 133, 139, 144
airlines xxiv, 37, 47, 48, 53, 72, 83, 84, 85, 86, 87, 88, 91, 92, 133
airplanes 15, 39
airports 8, 20, 39, 53, 54, 69, 71, 72, 74, 75, 76, 77, 83
airworthiness 15, 20, 39, 109, 110
algorithmic xviii, 7
algorithms xviii, xix, 3, 6, 7, 8, 9, 12, 14, 26, 36, 37, 38, 41, 43, 50, 51, 53, 69, 70, 71, 72, 73, 80, 81, 82, 88, 90, 93, 97, 98, 101, 103, 105, 109, 113, 117, 119, 121, 122, 123, 126, 127, 130, 135, 141, 143
artificial intelligence i, xviii, xix, 1, 3, 4, 5, 6, 33, 38, 53, 55, 58, 62, 116, 129, 135
ATC 8, 24, 29, 31, 32, 47, 124, 126, 127
ATCO 55, 57, 59, 60, 61, 64, 65, 66, 67, 68, 74, 75, 76, 77, 132
ATM 8, 20, 23, 77
Automated Decision Making 121
automation 10, 16, 47, 57, 59, 71, 73, 79, 91, 108, 109, 121, 122, 123, 124, 125, 126, 127, 130, 137, 138, 139, 140
autonomous xviii, xx, xxi, 3, 6, 11, 23, 26, 29, 30, 37, 45, 49, 52, 54, 60, 63, 73, 76, 103, 105, 120, 135, 136, 139, 141, 144
autonomous vehicles 7, 49, 50, 51, 52, 116, 117, 119, 120, 122, 124, 131, 132, 141, 142

aviation i, xviii, xxi, xxii, xxiii, xxiv, xxv, 1, 3, 5, 7, 8, 9, 10, 11, 13, 15, 16, 17, 19, 20, 21, 22, 23, 24, 25, 26, 28, 29, 31, 32, 34, 35, 38, 39, 40, 41, 44, 45, 46, 48, 53, 54, 55, 56, 57, 58, 59, 60, 63, 65, 68, 71, 75, 77, 81, 84, 91, 92, 93, 97, 102, 107, 108, 109, 112, 114, 115, 117, 119, 121, 123, 125, 127, 128, 129, 132, 133, 134, 138, 141, 142, 143, 144

Bayesian 21, 42, 123
bicyclist 15
Boeing 98, 103
business xxi, 13, 16, 33, 34, 37, 69, 70, 71, 78, 79, 87, 88, 108, 110, 121, 122

CBTA xxiv, 24, 25
chatbot xix, 69, 83, 94
climate xx, 54
cognitive 9, 10, 32, 45, 61, 75, 76, 101, 106, 111, 121, 137
complex 10, 13, 26, 35, 39, 40, 43, 44, 45, 48, 53, 54, 62, 84, 85, 86, 91, 93, 94, 108, 117, 133, 136, 137, 143
complexities xviii, 106
computer xviii, xix, 3, 5, 10, 26, 27, 28, 29, 37, 38, 41, 44, 45, 46, 52, 77, 85, 93, 98, 101, 108, 109, 111, 120, 123, 124, 130, 139
cultural xx, 40, 97
cybersecurity i

data xix, xx, 1, 5, 7, 8, 10, 12, 13, 15, 16, 17, 19, 22, 26, 27, 29, 31, 33, 35, 36, 37, 38, 39, 40, 41, 42, 43, 44, 45, 46, 47, 48, 50, 51, 52, 53, 54, 57, 59, 61, 62, 63, 64, 65, 67, 68, 69, 70, 71, 72, 73, 75, 78, 79, 80, 81, 82, 83, 84, 86, 87, 88, 89, 90, 91, 93, 94, 95, 96, 97, 98, 100, 101, 102, 103, 104, 105, 106, 108, 109, 110, 111, 112, 113, 116, 117, 120, 121, 122, 123, 124, 125, 126, 127, 128, 130, 131, 133, 134, 135, 136, 141, 142, 143
Data-Driven Decision Automation 122
decision i, xx, xxi, 1, 10, 11, 17, 23, 24, 25, 28, 30, 31, 32, 35, 41, 42, 43, 45, 46, 47, 48, 50, 54, 56, 59, 61, 67, 79, 88, 94, 96, 97, 99, 100, 101, 104, 105, 106, 108, 109, 111, 116, 117, 121, 122, 123, 124, 125, 127, 130, 132, 133, 135, 137, 139, 143

decision-makers i, 123
decision-making 121
decision support systems 123
Digital Twins 39, 43, 113
drones 3, 20, 45, 53, 54, 58, 59, 60, 143

EAAP xxiv, xxv, 1
EASA xxiv, xxv, 1, 3, 5, 8, 15, 19, 20, 21, 22, 23, 44, 48, 57, 58, 60, 75, 77, 129, 135
EATEO 1
economy xix, 12, 78
ecosystem xxi, xxiii, 6, 47, 54, 71, 79, 108, 128
efficiency xviii, xx, 6, 8, 9, 10, 11, 26, 29, 31, 36, 37, 38, 39, 41, 42, 43, 45, 46, 53, 69, 70, 71, 72, 73, 74, 79, 81, 82, 84, 91, 94, 96, 103, 104, 109, 115, 117, 118, 119, 120, 121, 122, 123, 124, 126, 131, 133, 136, 137, 141, 142, 143, 144
empirical xviii
engineering i, xviii, 26, 29, 31, 78, 81, 99, 101, 109, 113, 118, 119
environmentally 16
ergonomics xxv, 9, 10, 46, 58, 77, 135, 138, 140
ethical xviii, xx, xxi, xxii, 5, 16, 17, 18, 19, 22, 28, 49, 50, 52, 54, 72, 84, 90, 106, 116, 117, 120, 128, 129, 131, 132, 133, 135, 136, 141, 142, 144
ethics xviii, 5, 22, 26, 28, 49, 129, 135, 140, 142
EU 17, 18, 19, 20, 23, 102, 103, 128, 131
EU Commission 17, 19, 20
Europe xxi, xxv, 15, 17, 20, 23, 44, 50, 51, 60, 132
European Council 17
European Parliament 17

FAA 15, 48, 64, 65, 99, 109, 125, 126, 127
fuel 8, 13, 32, 36, 37, 38, 52, 53, 78, 79, 96, 102, 126

generation 12, 15, 139

hardware xxi, xxiv, 3, 27, 46, 127
human i, xviii, xix, xx, xxv, 1, 2, 3, 7, 9, 10, 12, 15, 19, 21, 22, 23, 31, 33, 34, 37, 41, 45, 50, 56, 57, 58, 59, 60, 62, 64, 66, 68, 72, 73, 75, 76, 77, 81, 82, 83, 84, 89, 91, 94, 97, 98, 99, 100, 101, 104, 107, 108, 109, 110, 113, 115, 116, 117, 118, 119, 120, 121, 123, 124, 129, 130, 132, 137, 138, 141, 142
human factors i, xxiv, xxv, 9, 46, 55, 58, 63, 65, 77, 115, 127, 138
human machine interaction 117, 119, 120, 124

human performance i, xviii, xx, 2, 4, 6, 8, 10, 12, 14, 16, 18, 20, 22, 24, 28, 30, 32, 34, 36, 38, 40, 42, 44, 46, 48, 50, 52, 54, 56, 58, 60, 62, 64, 66, 68, 70, 72, 74, 76, 78, 80, 82, 84, 86, 88, 90, 92, 94, 96, 98, 100, 102, 104, 106, 108, 110, 112, 114, 116, 118, 120, 122, 124, 126, 130, 132, 134, 136, 138, 140, 142, 144
human resources i, 33

industry xviii, xix, xx, xxi, 1, 2, 12, 13, 14, 16, 20, 22, 26, 27, 28, 29, 30, 31, 33, 38, 39, 41, 45, 49, 53, 70, 71, 78, 79, 81, 83, 84, 85, 87, 91, 96, 101, 103, 104, 105, 109, 117, 119, 121, 122, 128, 129, 130, 136, 137, 140, 142, 144
infrastructure xviii, 6, 8, 16, 36, 50, 57, 71, 73, 79, 80, 103, 141, 144
interaction xviii, 9, 10, 57, 60, 75, 84, 106, 115, 117
Internet of Things 54, 93, 103, 132, 142

land i, 13, 15
learning xix, 7, 12, 13, 14, 15, 26, 27, 28, 30, 31, 32, 34, 35, 39, 40, 41, 42, 44, 45, 47, 48, 53, 55, 56, 61, 64, 67, 71, 72, 73, 74, 76, 79, 83, 89, 90, 93, 95, 97, 99, 108, 109, 110, 113, 117, 118, 122, 123, 124, 130, 141, 142, 143
legal 17, 49, 52, 65, 72, 81, 129, 131, 132, 133, 141, 142
legislation 1, 16, 17, 64, 78, 131, 132, 136, 138
logistics i, xxii, 6, 8, 11, 13, 26, 29, 36, 69, 96, 106, 139, 141

machine 3, 4, 5, 39, 50, 57, 59, 93, 98, 108, 113, 137
machine learning xix, 4, 14, 20, 44, 63, 82, 83, 93, 123, 129, 135
maintenance xix, xx, 6, 8, 15, 31, 32, 33, 34, 37, 38, 39, 40, 41, 42, 43, 44, 46, 70, 73, 94, 95, 96, 97, 98, 99, 100, 101, 102, 103, 104, 105, 108, 109, 110, 111, 112, 113, 116, 133, 135, 136, 139, 141, 142, 144
management i, xviii, xix, xx, xxi, 1, 2, 6, 7, 8, 11, 13, 16, 19, 20, 26, 31, 33, 34, 43, 46, 47, 48, 51, 52, 53, 54, 55, 61, 62, 63, 64, 65, 66, 68, 75, 76, 78, 81, 82, 85, 86, 87, 97, 98, 101, 102, 103, 104, 105, 109, 121, 122, 124, 127, 129, 130, 132, 133, 136, 137, 139, 141, 142, 143, 144

Index

manufacturing xix, 118
maritime i, xxii, 13, 14, 33, 35, 37, 65, 70, 71, 77, 78, 79, 80, 81, 104
methodology 13, 43, 64, 102, 107, 119
methods xix, 24, 29, 30, 35, 39, 40, 41, 42, 49, 50, 62, 63, 65, 87, 93, 117

NASA 43, 102
natural language xix, 30, 39, 42, 83, 123
natural language processing 46, 63, 89, 93, 107
neural networks 38, 40, 93, 102, 123

online training i, xxi
operations i, xviii, xix, xx, xxi, 7, 10, 13, 14, 16, 20, 23, 24, 31, 32, 33, 34, 36, 37, 38, 39, 40, 42, 43, 45, 46, 51, 53, 55, 57, 58, 59, 64, 65, 66, 68, 70, 71, 73, 74, 76, 77, 78, 79, 80, 81, 82, 85, 89, 95, 97, 101, 102, 103, 105, 106, 109, 113, 117, 121, 124, 125, 127, 137, 138, 141, 142

passengers 8, 15, 16, 33, 72, 84, 87, 88
PdM 94, 98, 101, 102, 103, 104, 105, 110, 111, 112, 113
performance xviii, xix, xx, 1, 10, 13, 14, 24, 31, 32, 33, 37, 40, 41, 42, 46, 48, 61, 63, 64, 66, 67, 68, 76, 77, 78, 83, 94, 95, 101, 105, 106, 109, 110, 115, 124, 126, 130, 137, 138, 140
pilots 10, 24, 32, 45, 46, 47, 48, 55, 56, 57, 58, 63, 66, 68, 94, 124, 132
policymakers xviii, xxii, 3, 12, 14, 136
practitioners xxii, 100
predictive 8, 29, 35, 36, 37, 38, 39, 40, 41, 42, 43, 44, 46, 53, 62, 73, 79, 86, 90, 94, 96, 97, 100, 101, 102, 104, 105, 110, 111, 124, 134, 135, 136, 141, 142, 143, 144
processes xx, xxiv, 13, 31, 38, 39, 43, 45, 56, 72, 74, 75, 78, 81, 86, 88, 95, 101, 105, 108, 110, 117, 121, 122, 123, 124, 127, 133, 135, 136, 139, 143
professionals i, xviii, 14, 26, 45
psychology xviii, 138
public xix, 5, 6, 8, 12, 13, 14, 15, 16, 19, 41, 50, 86, 90, 115, 121, 122, 128, 142, 144
Purdue 1, 29

railways xix, xx, 8, 15, 18, 19, 15, 80, 81, 82, 103
regulator 17

regulatory xxii, 1, 5, 14, 15, 16, 17, 18, 19, 20, 54, 55, 72, 95, 108, 127, 131, 132, 135, 136, 143, 144
reliability 12, 13, 23, 38, 39, 41, 42, 43, 50, 54, 81, 82, 95, 96, 101, 102, 103, 109, 110, 115, 116, 120, 121, 122, 140, 142
researchers xxii, 30, 119, 144
resource 1, 33, 42, 44, 48, 72, 75, 91, 143
risk xix, 7, 15, 16, 17, 18, 19, 20, 22, 23, 31, 32, 37, 38, 46, 57, 58, 62, 63, 64, 66, 76, 95, 96, 101, 106, 121, 132, 133, 136, 143, 144
road 8, 15, 36, 49, 50, 51, 52, 131
robotics xix, 30, 105, 121
Rule-Based Decision Automation 121, 122

safety i, xviii, xix, xx, xxi, xxii, xxiv, xxv, 1, 3, 5, 6, 7, 8, 10, 11, 13, 14, 16, 17, 18, 19, 20, 22, 23, 26, 28, 29, 31, 33, 34, 36, 37, 38, 39, 41, 43, 44, 45, 46, 49, 50, 54, 55, 56, 57, 58, 59, 60, 62, 63, 64, 65, 66, 67, 70, 74, 75, 76, 77, 78, 79, 80, 81, 96, 97, 100, 101, 102, 103, 107, 110, 115, 116, 117, 118, 119, 120, 122, 124, 127, 128, 131, 132, 133, 134, 136, 137, 138, 139, 140, 141, 142, 143, 144
Safety Management System 33, 62
science xviii, 2, 15, 26, 27, 29, 31, 90, 104, 108, 109, 111, 123, 128, 132
sea i, 13
security i, xx, xxi, 8, 16, 17, 18, 19, 20, 22, 34, 38, 51, 54, 69, 70, 71, 72, 73, 74, 79, 84, 91, 105, 109, 116, 121, 131, 136
security management 33
SESAR 125
shipping companies 34, 37, 70, 77, 79
simulation xxiv, 24, 32, 36, 61, 119, 120
smart 3, 4, 5, 26, 30, 49, 52, 63, 71, 79, 103, 105, 116
social xviii, xxi, 1, 3, 5, 12, 18, 49, 50, 51, 52, 54, 88, 90, 144
software xxi, 1, 3, 15, 22, 23, 27, 28, 45, 51, 99, 101, 105, 109, 112, 127
specialists i
stakeholders xxi, 1, 11, 12, 14, 20, 26, 27, 54, 72, 82, 117, 129, 136, 140, 143, 144
statistical 22, 35, 62, 63, 123
supply chain 8, 36, 39, 71, 79, 95, 96, 105, 106, 130
sustainability xviii, 11, 29, 37, 102, 115, 131, 136, 141, 144

technology i, xviii, xix, xx, xxi, xxiii, xxiv, 1–6, 8–12, 14, 16, 17, 20, 23–30, 32, 36, 38, 41, 45, 47–50, 52–55, 57–59, 66, 71–73, 79–83, 85, 88–91, 99–105, 107–112, 115–120, 122–124, 128–132, 135–139, 141, 143, 144
traffic xviii, 6, 7, 8, 11, 15, 20, 26, 29, 30, 36, 45, 46, 49, 50, 51, 52, 53, 54, 55, 58, 62, 64, 65, 66, 67, 68, 74, 75, 76, 80, 81, 82, 96, 104, 105, 117, 119, 121, 122, 124, 125, 127, 137, 142, 143, 144
training xxi, xxiv, xxv, 1, 14, 24, 25, 26, 27, 29, 31, 32, 33, 34, 35, 37, 38, 39, 41, 43, 45, 46, 47, 95, 110, 116
trains xix, xx, 80, 81, 103, 122
transportation i, xviii, xix, xx, xxi, xxii, xxiii, 1, 2, 3, 4, 6, 7, 8, 11, 12, 13, 14, 15, 16, 26, 27, 29, 30, 31, 36, 37, 49, 52, 66, 68, 69, 71, 72, 79, 80, 81, 82, 83, 96, 101, 102, 103, 104, 105, 106, 115, 116, 117, 118, 119, 120, 121, 122, 131, 132, 135, 136, 137, 140, 141, 142, 143, 144
trust xx, 1, 11, 12, 14, 22, 40, 50, 51, 55, 68, 75, 88, 115, 116, 119, 120, 128, 130, 135

United States xxv, 16, 29, 131, 132
urban planning xviii

vehicles 4, 6, 7, 8, 43, 45, 49, 50, 51, 52, 58, 116, 117, 120, 135, 141, 142, 144
virtual reality 31, 34
visual recognition 93